THIS HUMAN ANATOMY COLORING BOOK

I0480897

BELONG TO

_____

_____

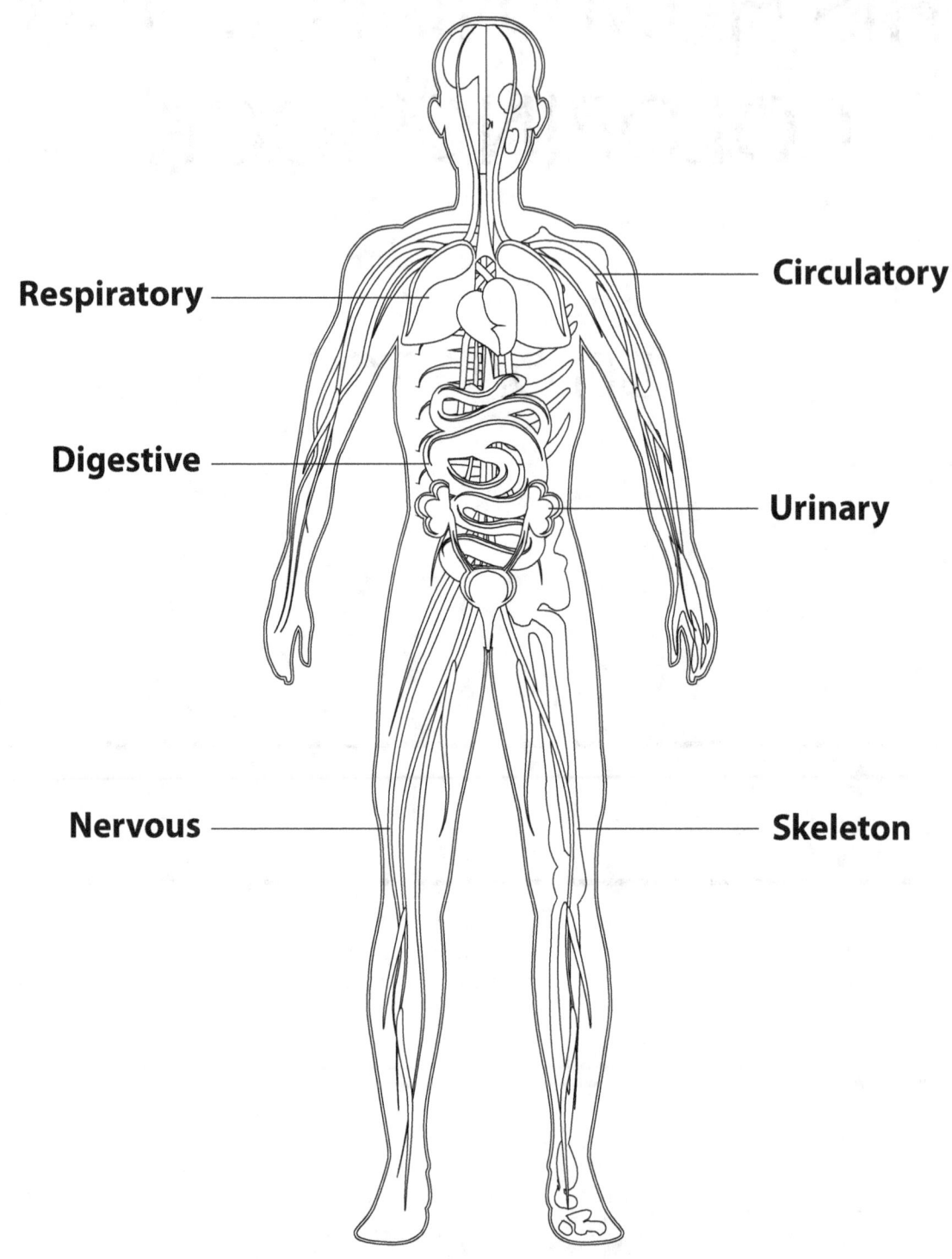

Respiratory

Circulatory

Digestive

Urinary

Nervous

Skeleton

Human Anatomy

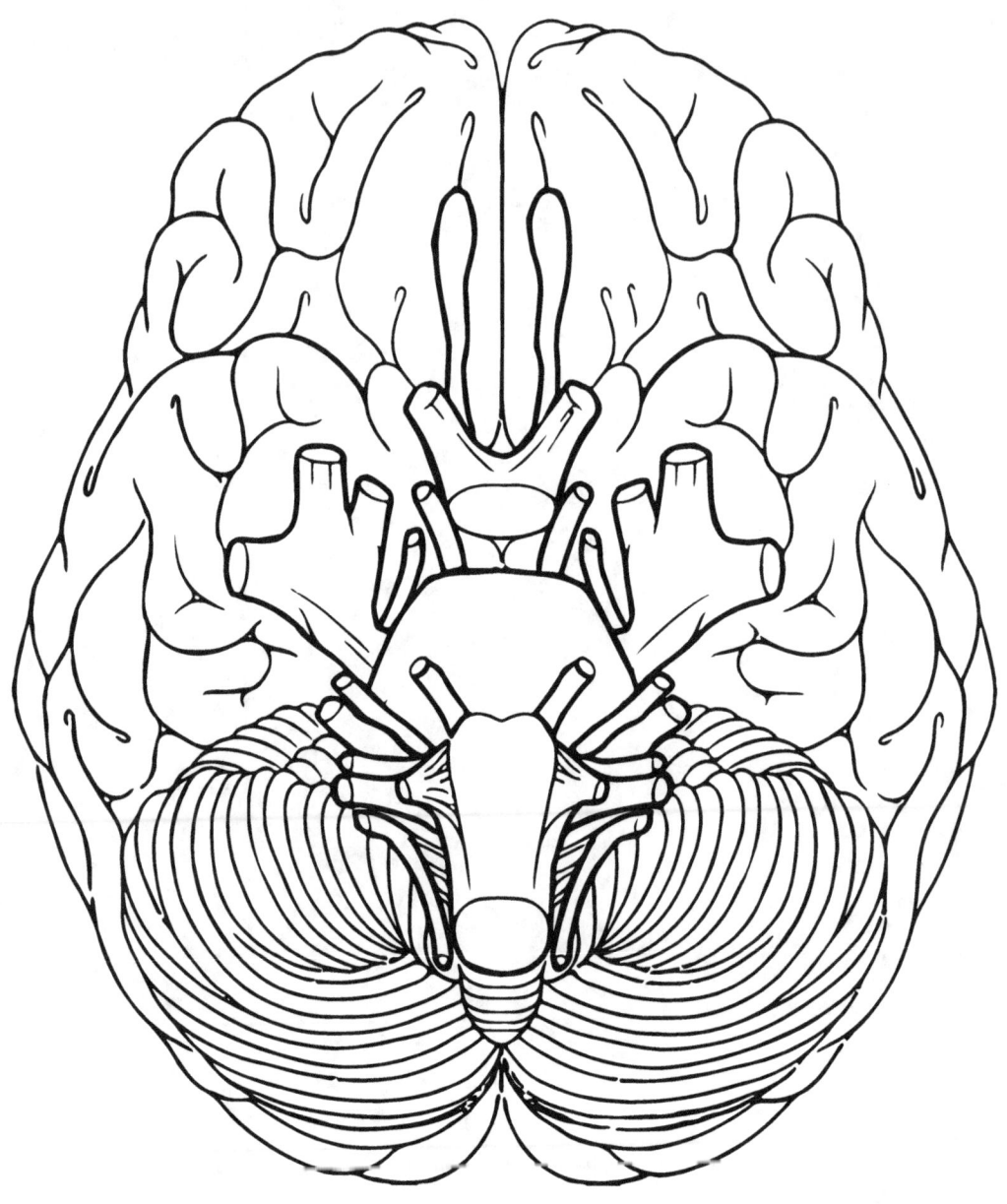

Brain

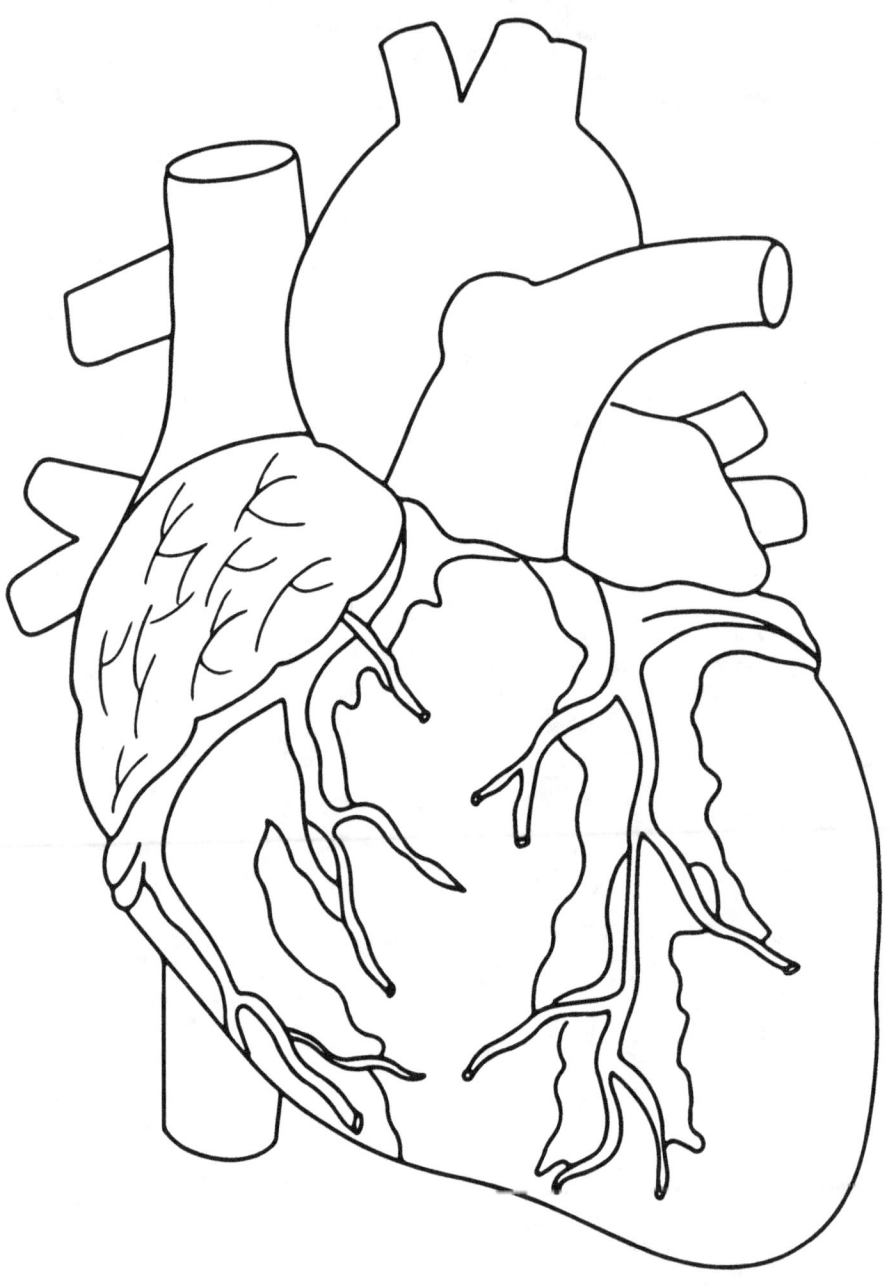

Heart

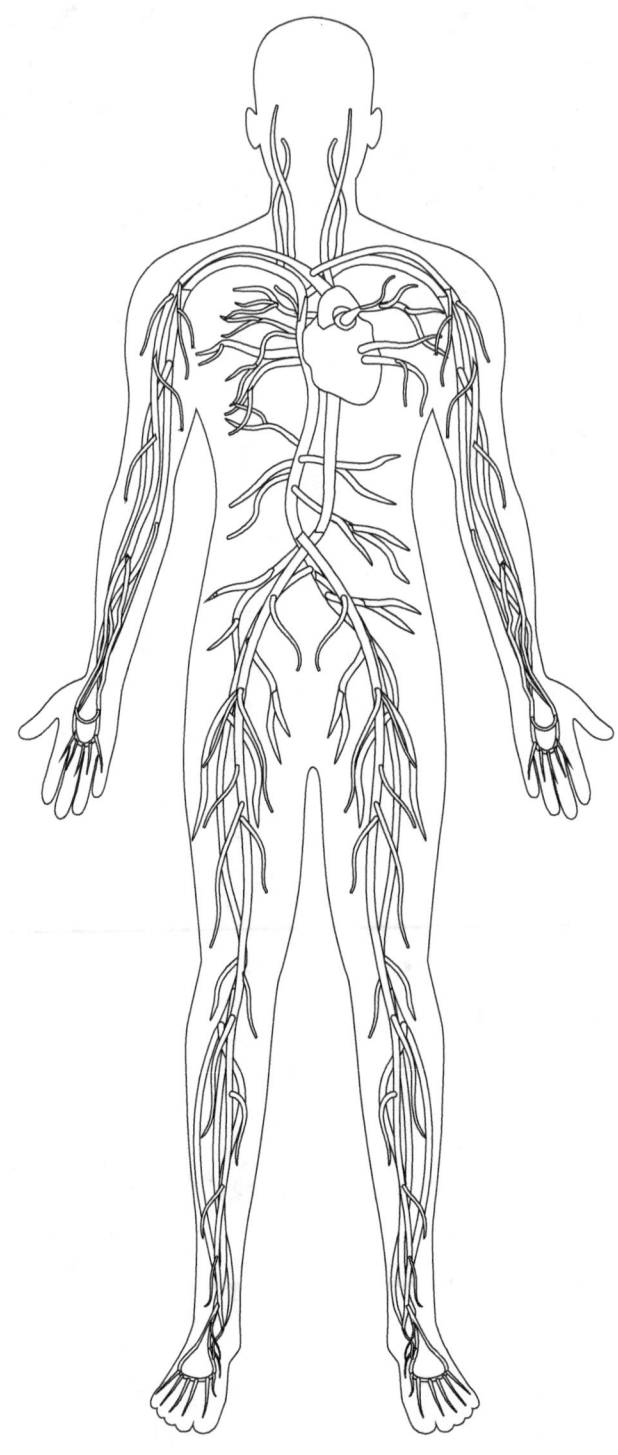

Circulatory System

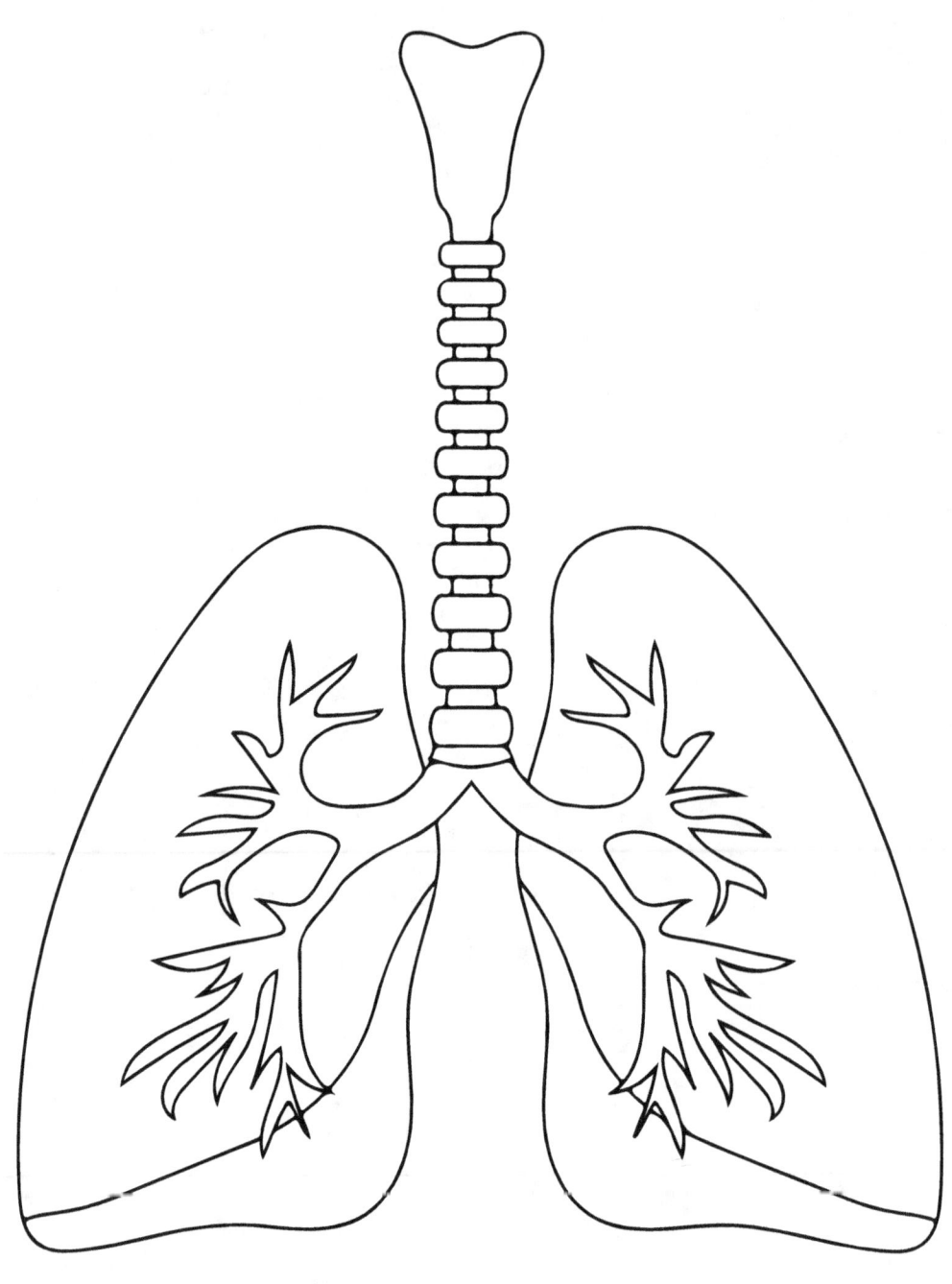

Lungs

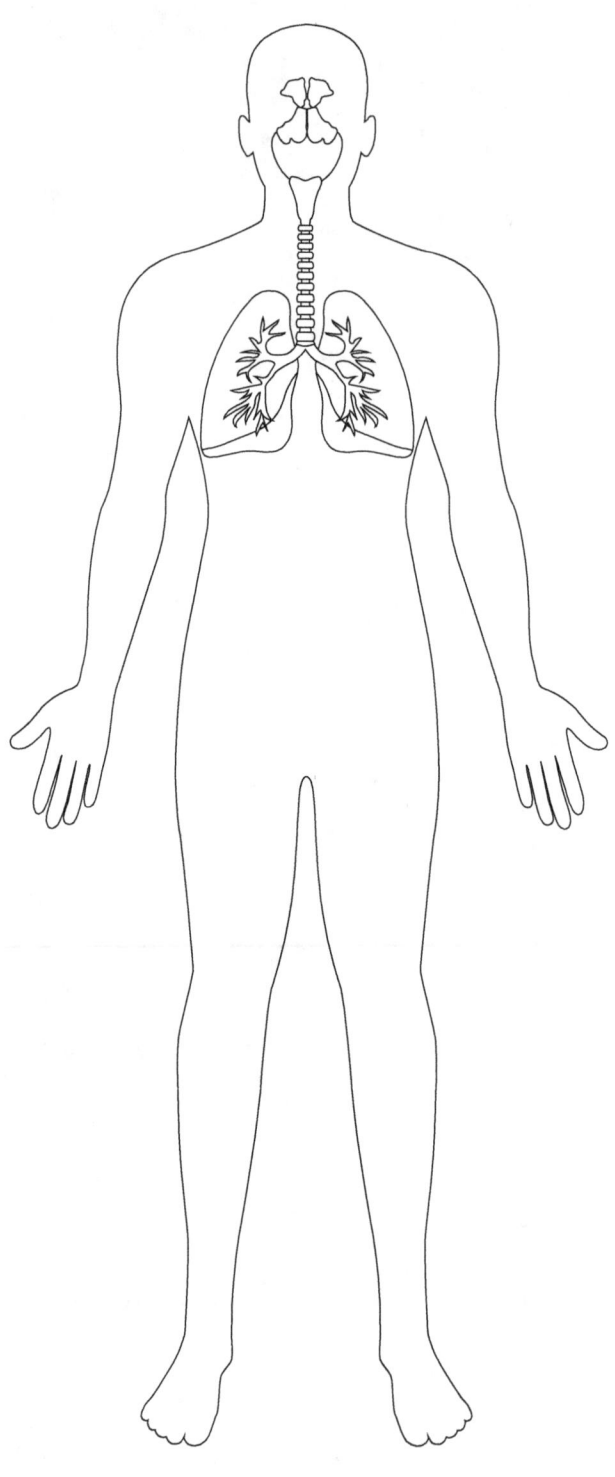

Respiratory System

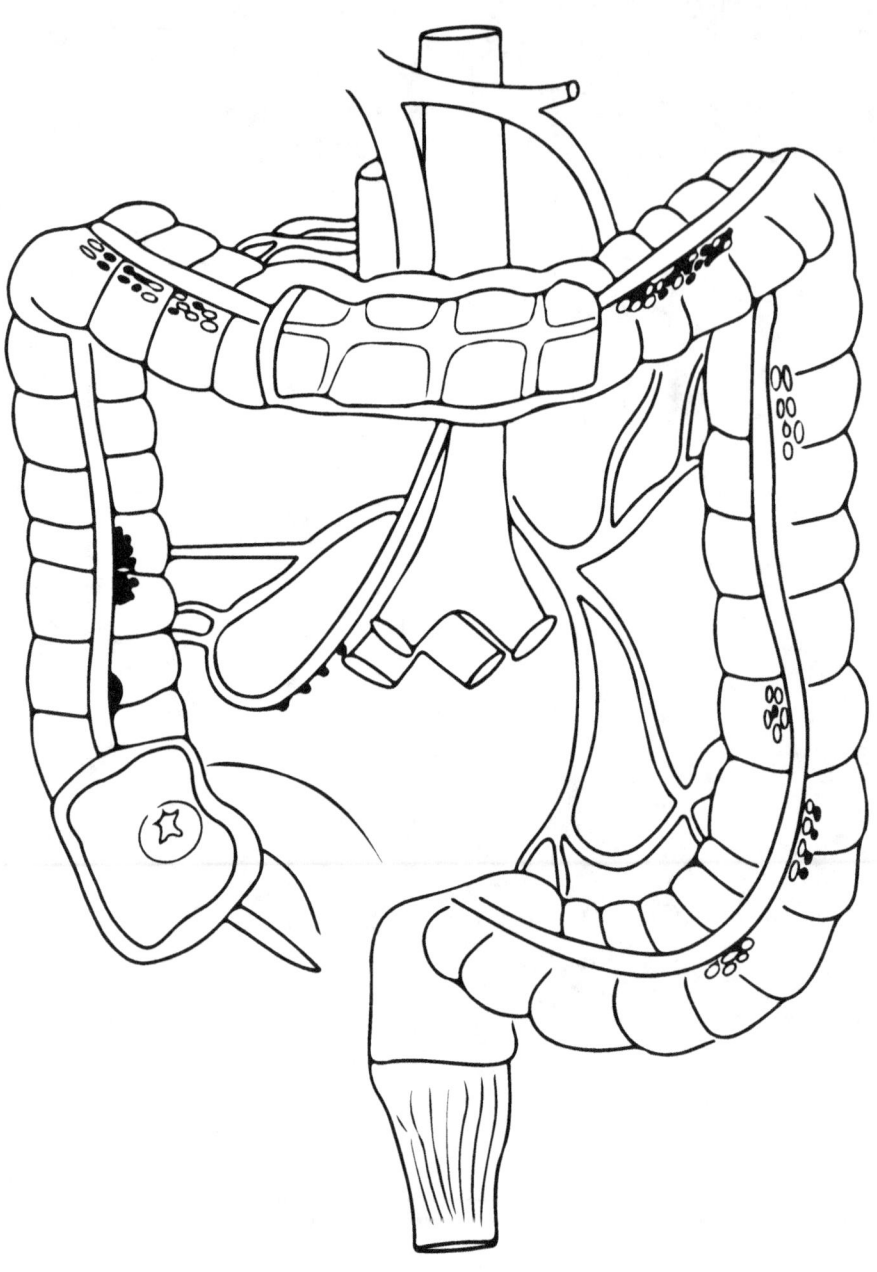

Intestine

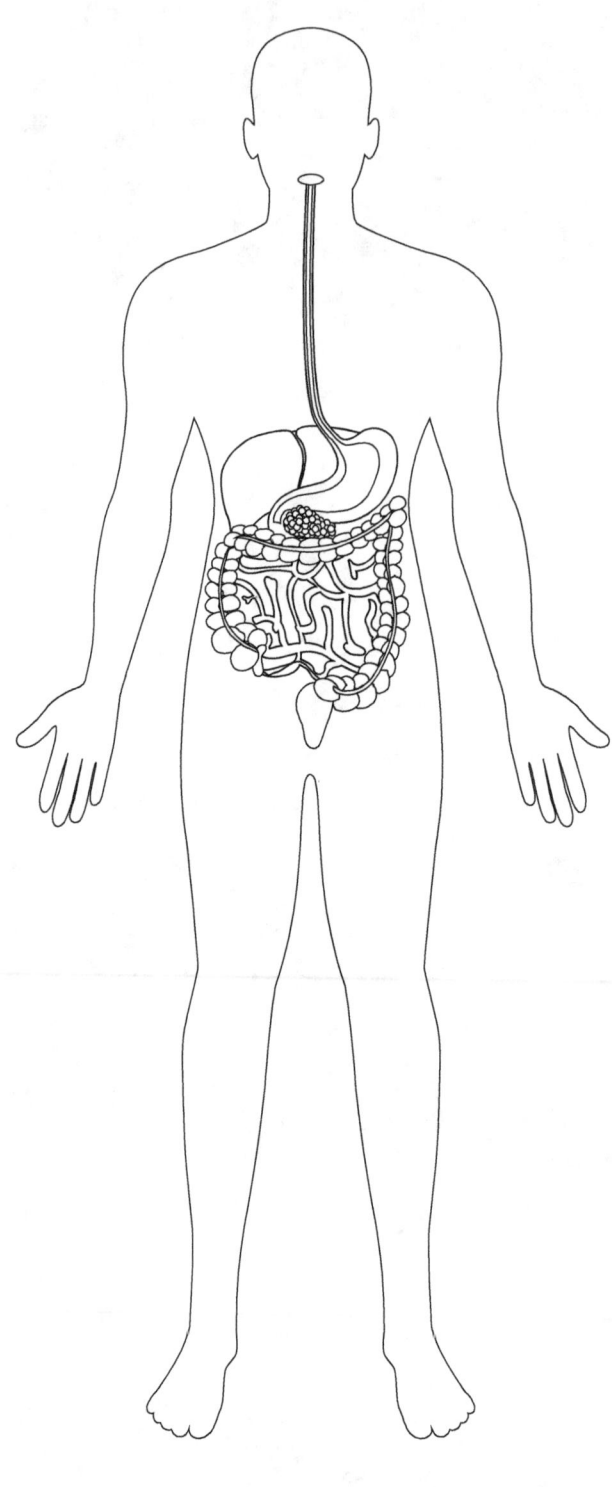

Digestive System

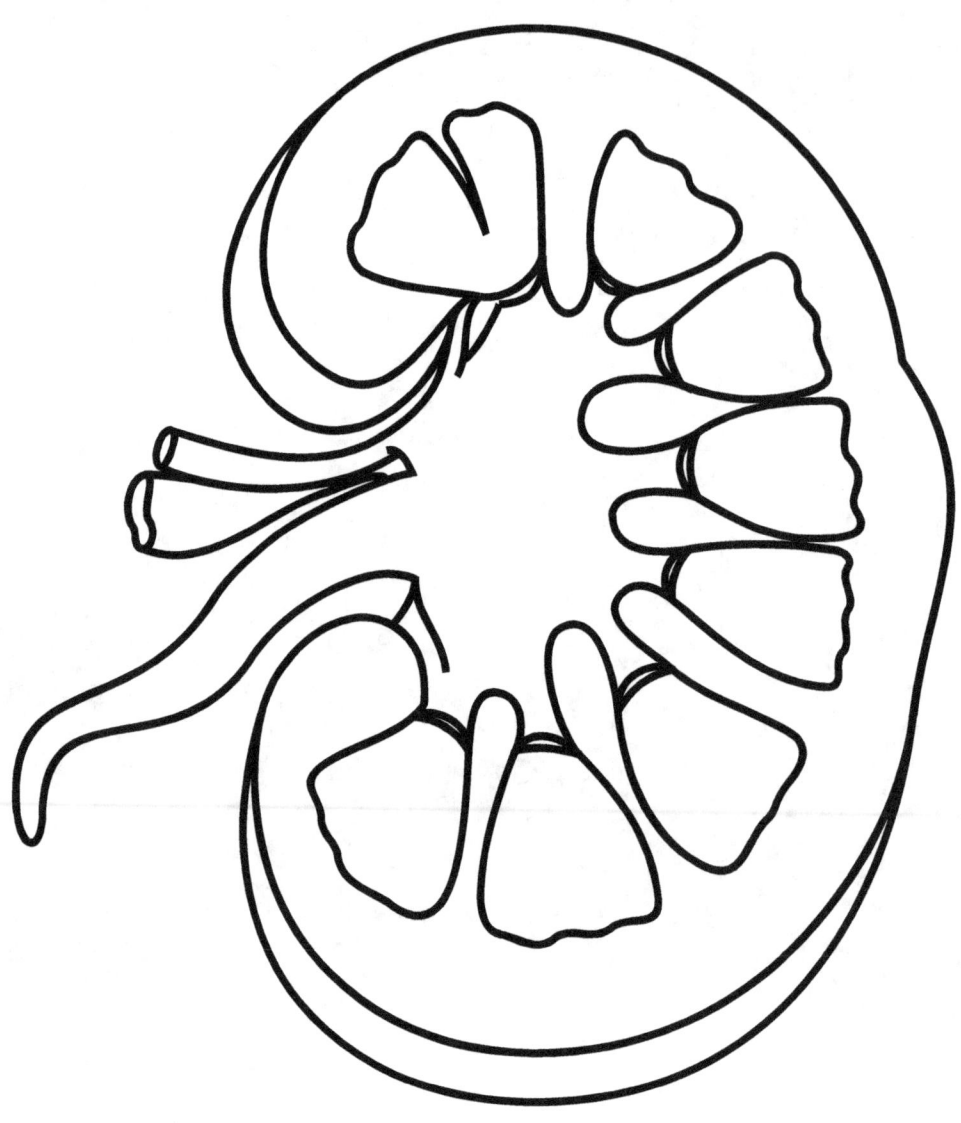

Kidney

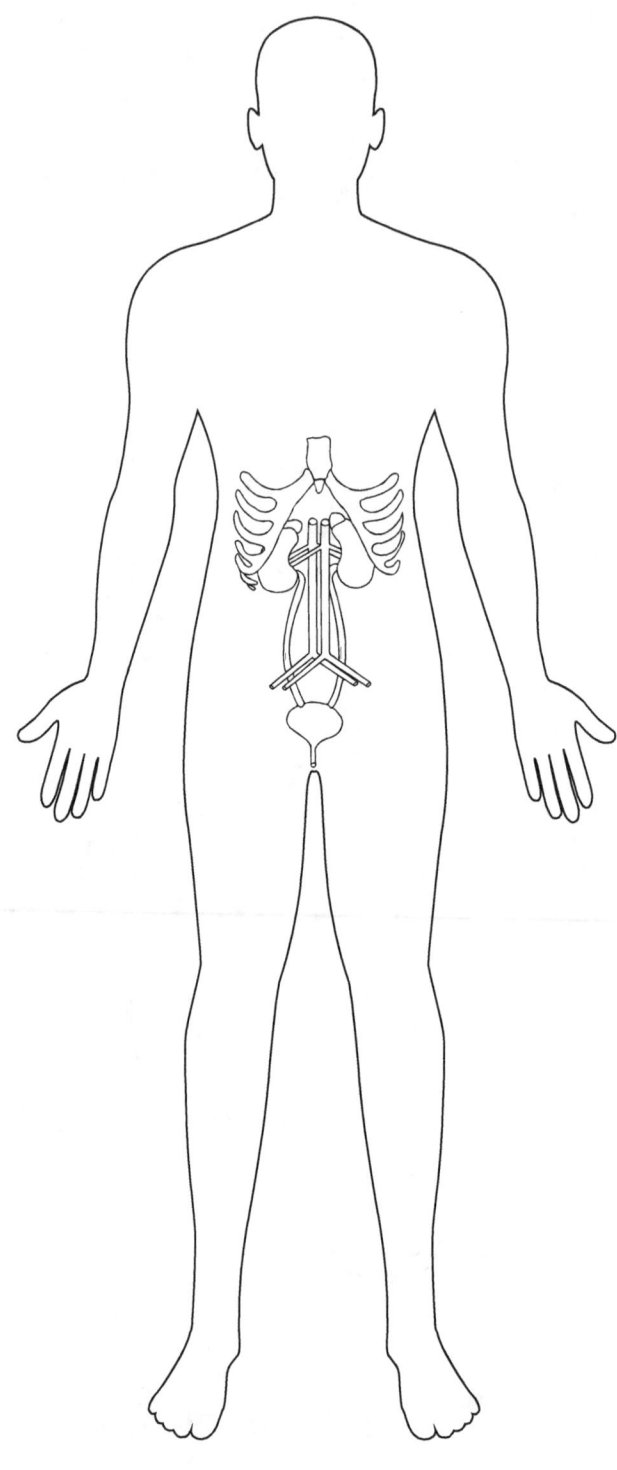

Urinary System

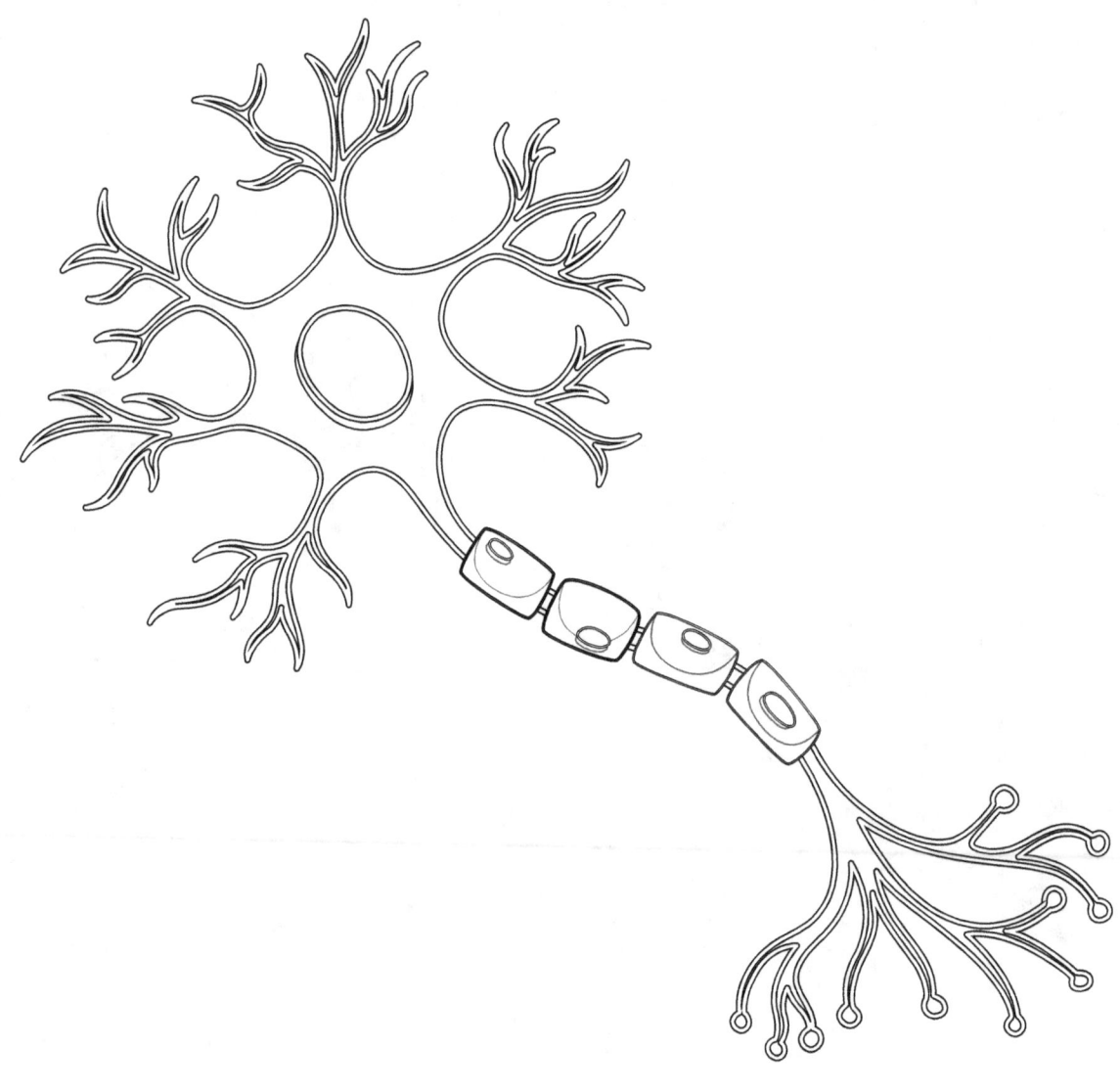

Nerve Cell

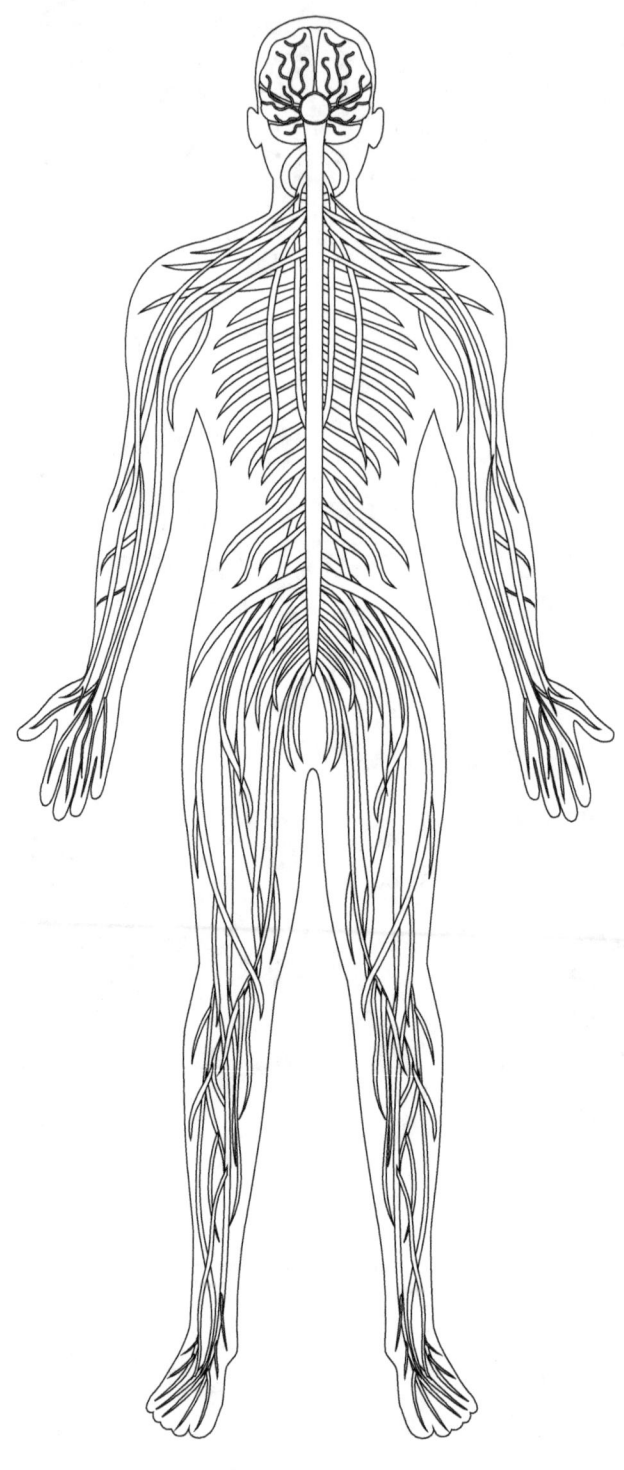

Nervous System

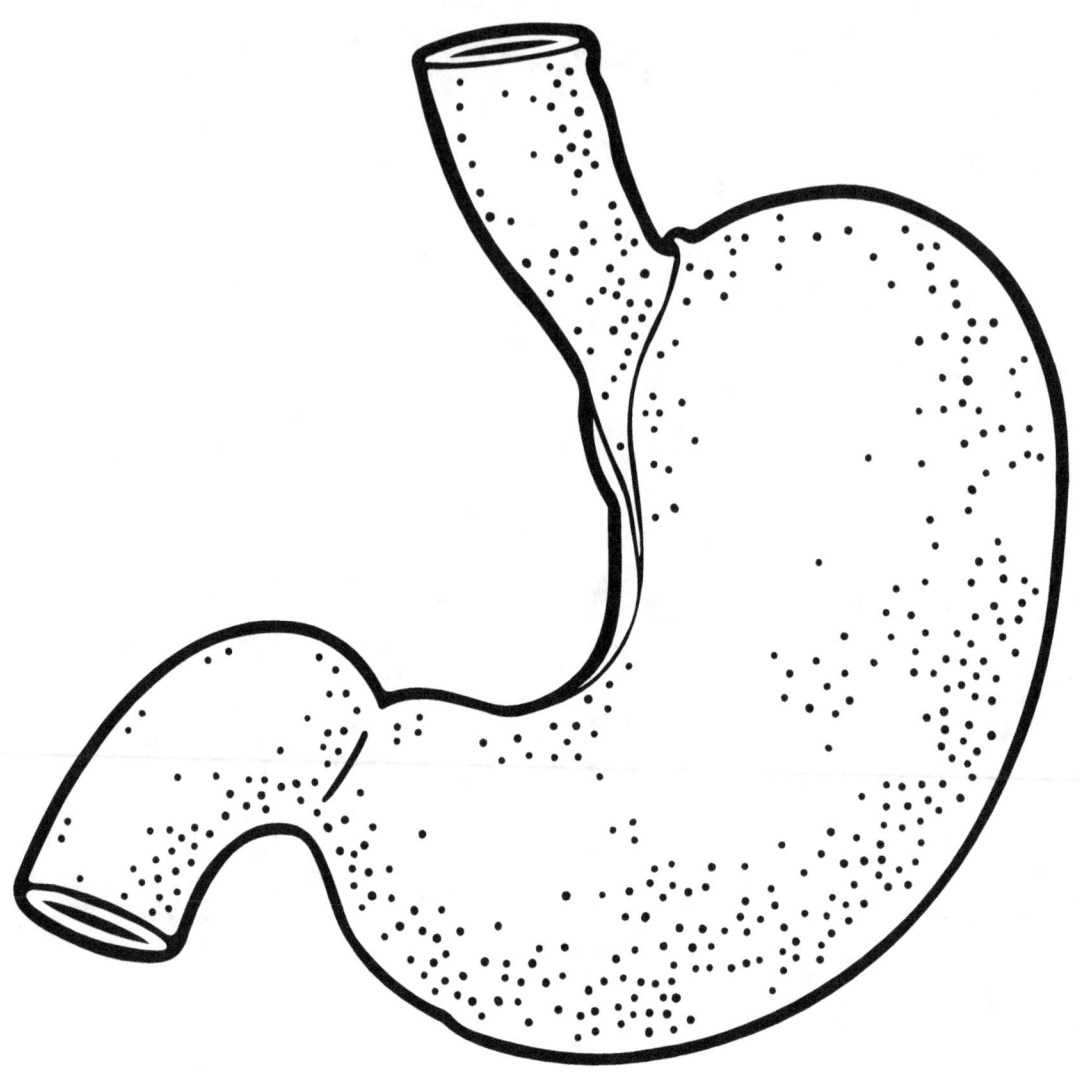

Stomach

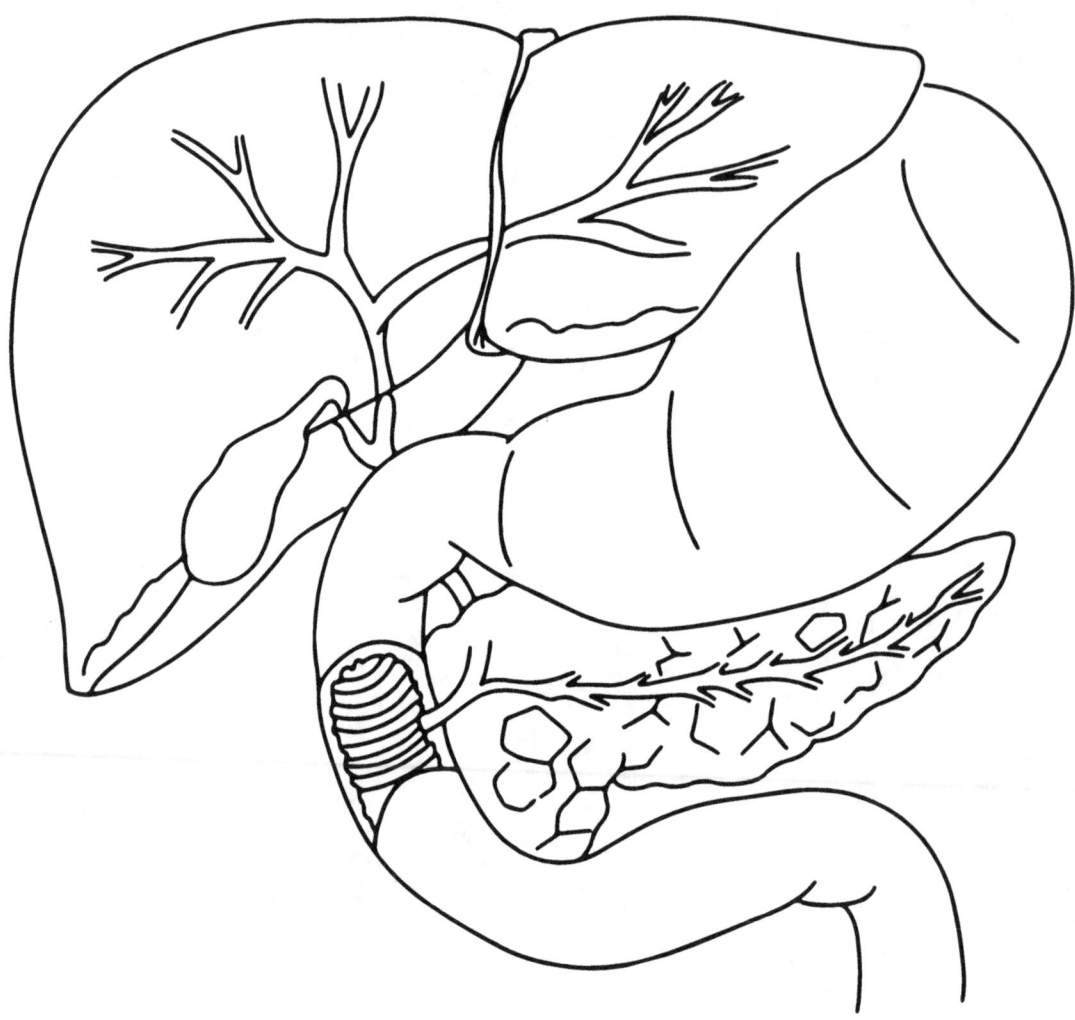

Liver

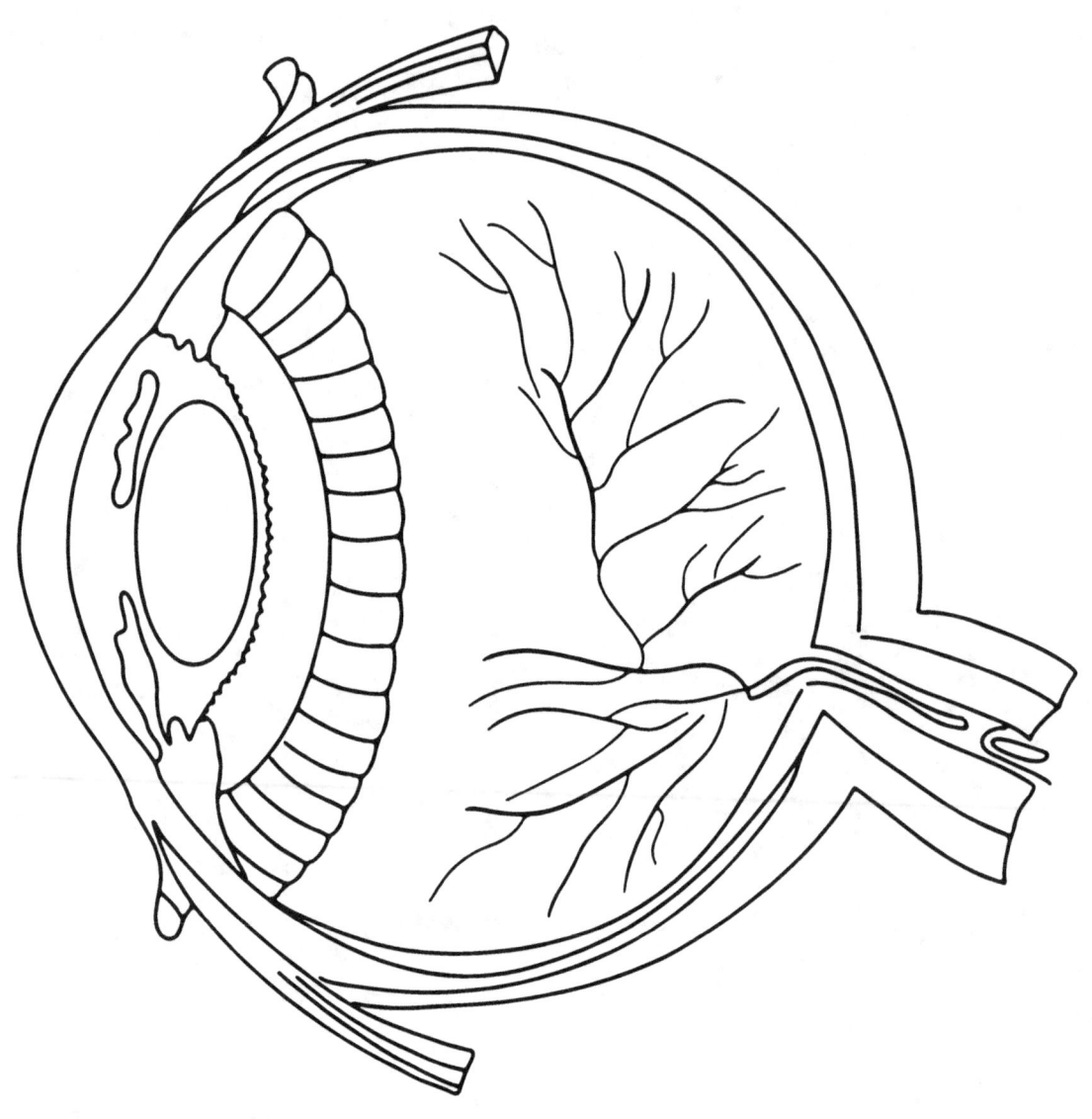

Eye

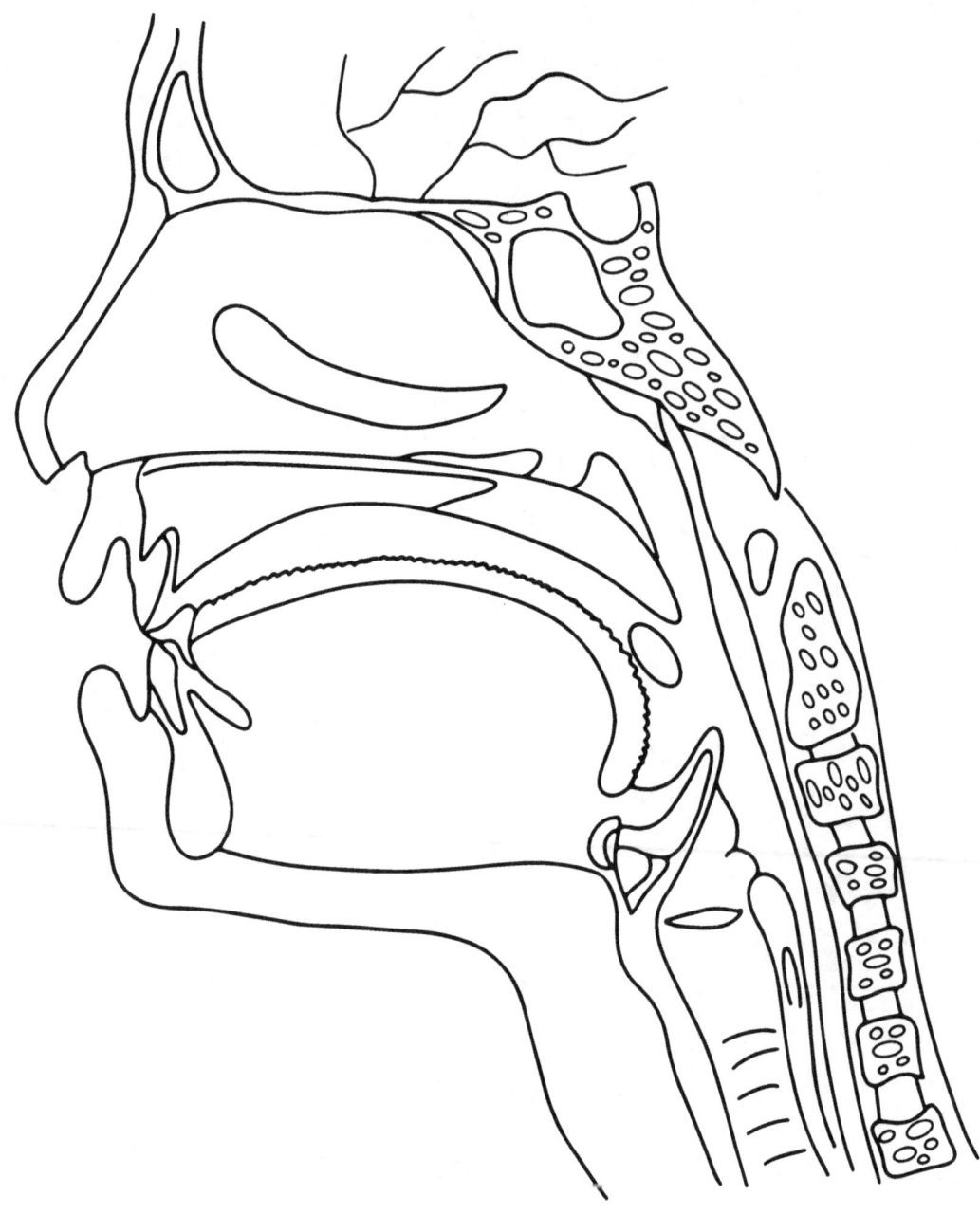

Nose and Throat

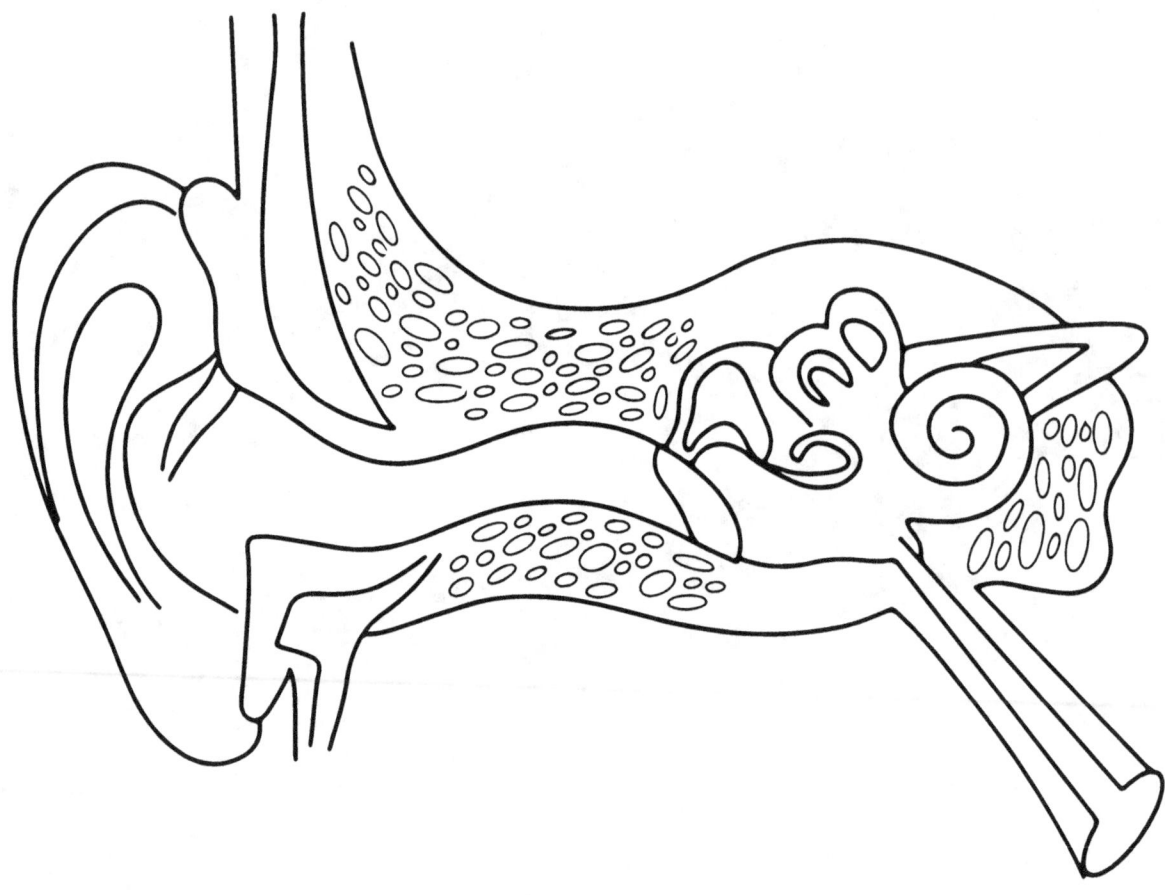

Ear

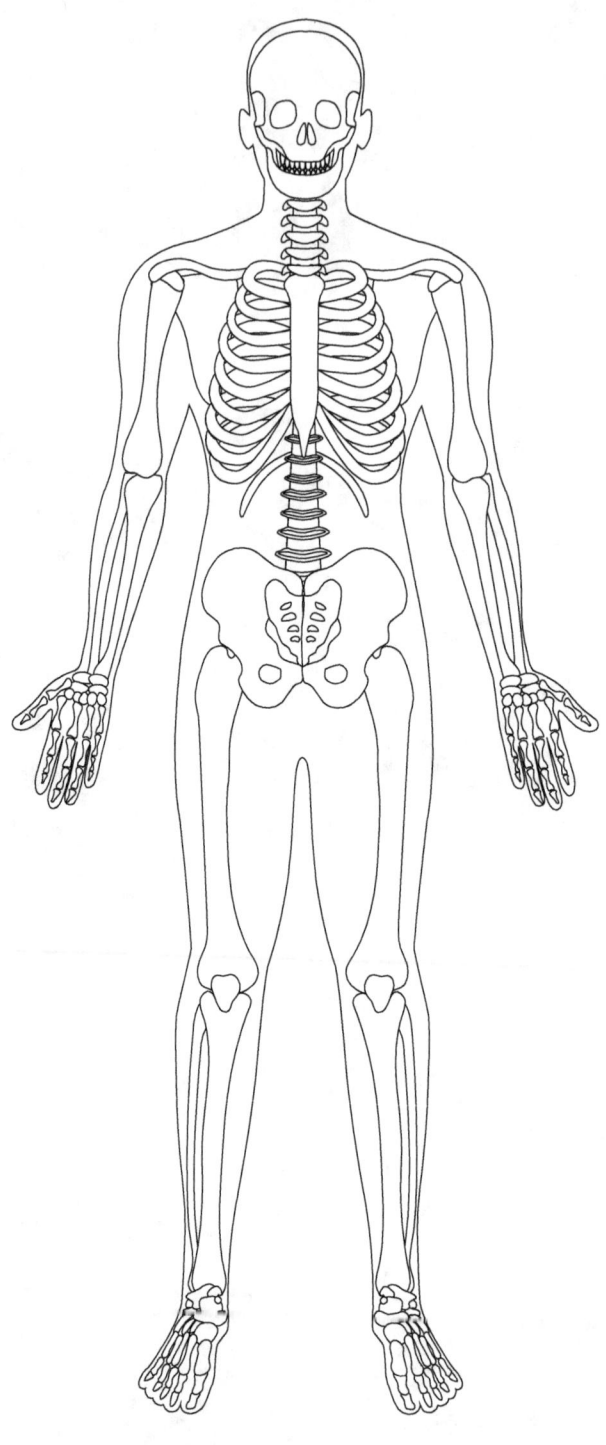

Skeletal System

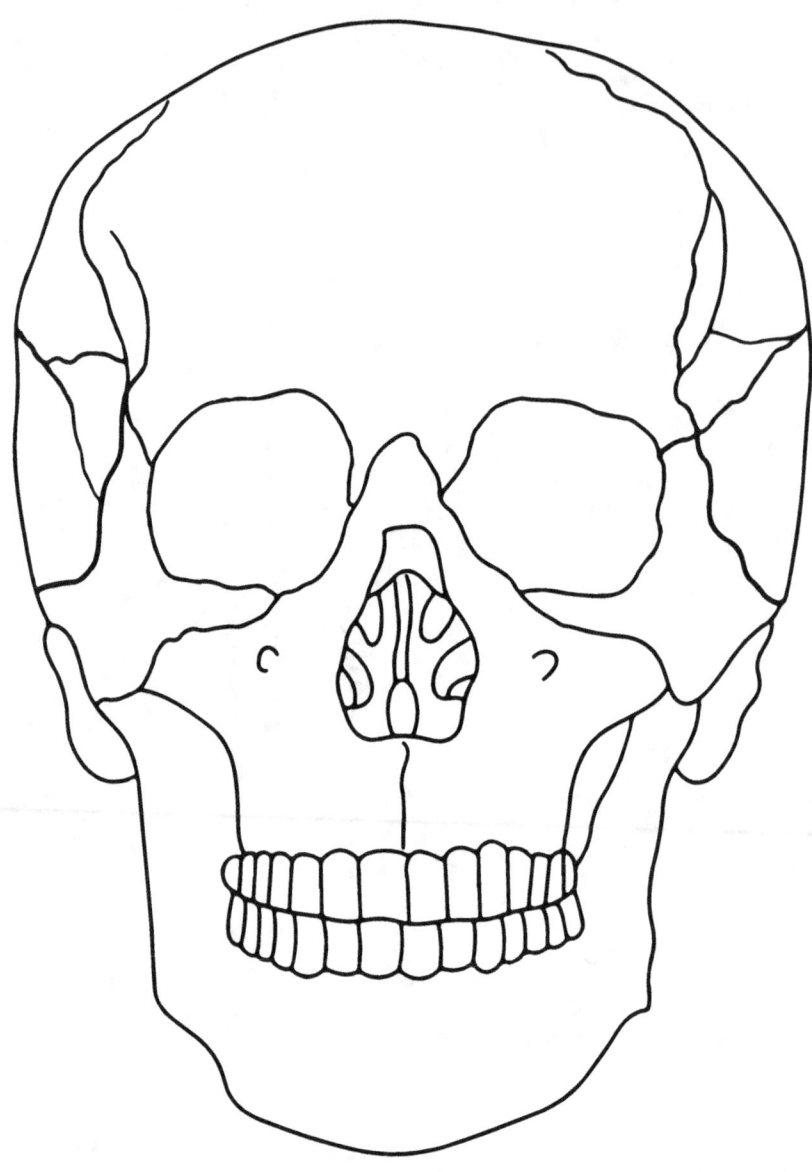

Skull

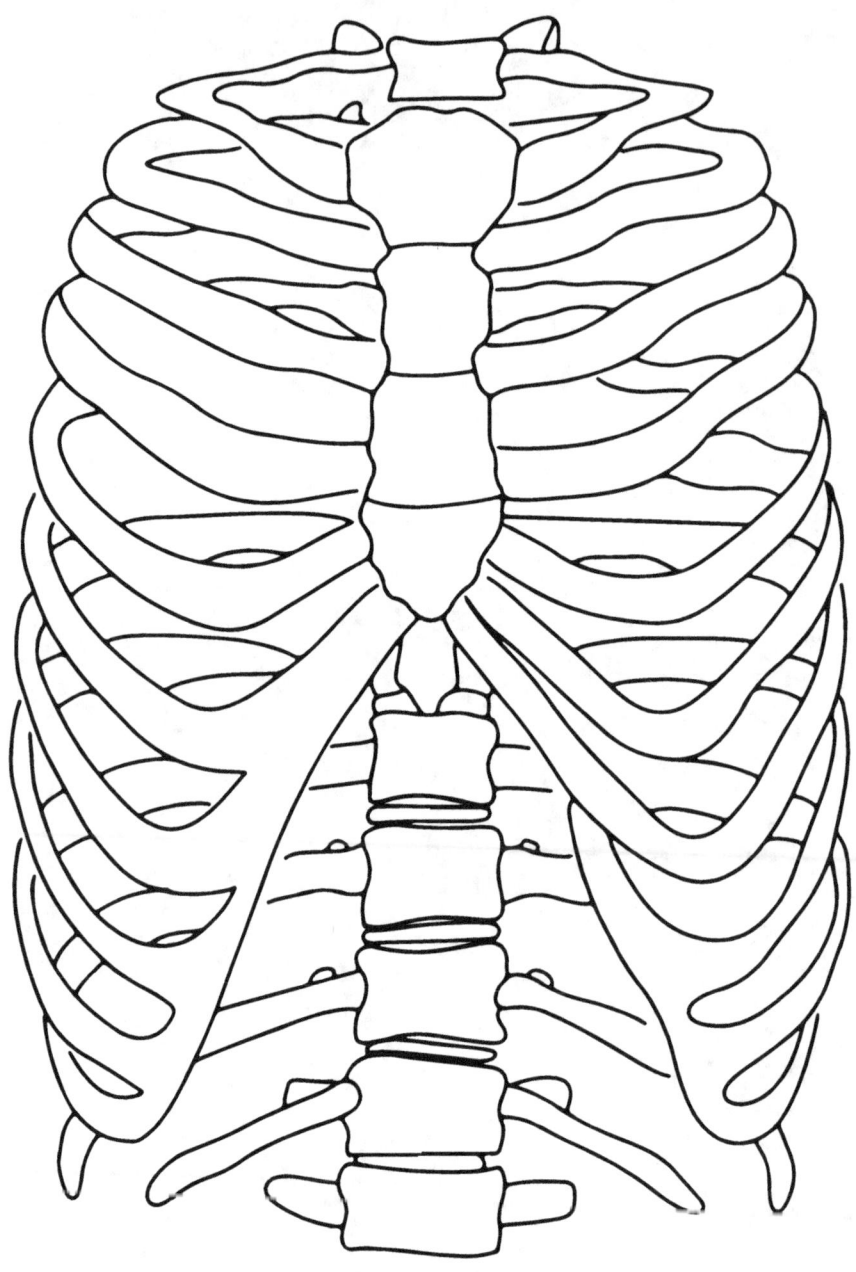

Ribs

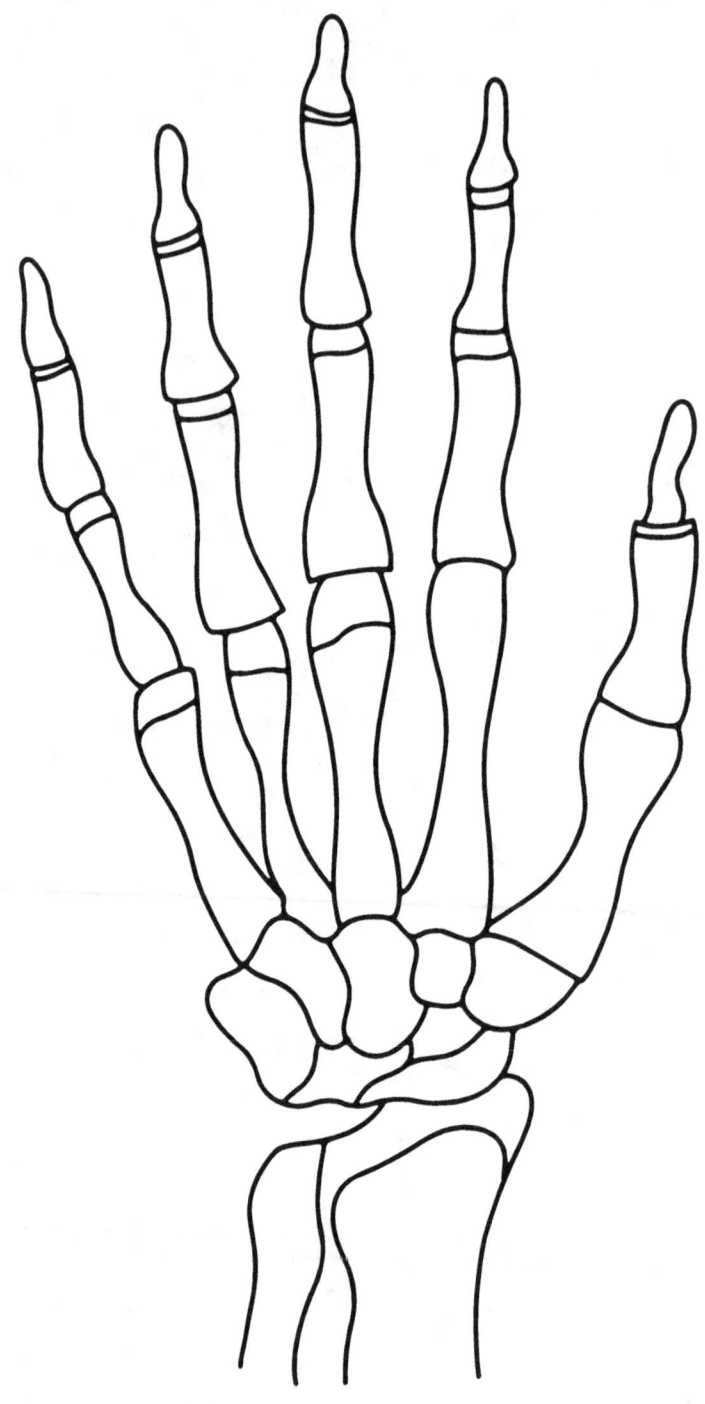

Hand and Wrist

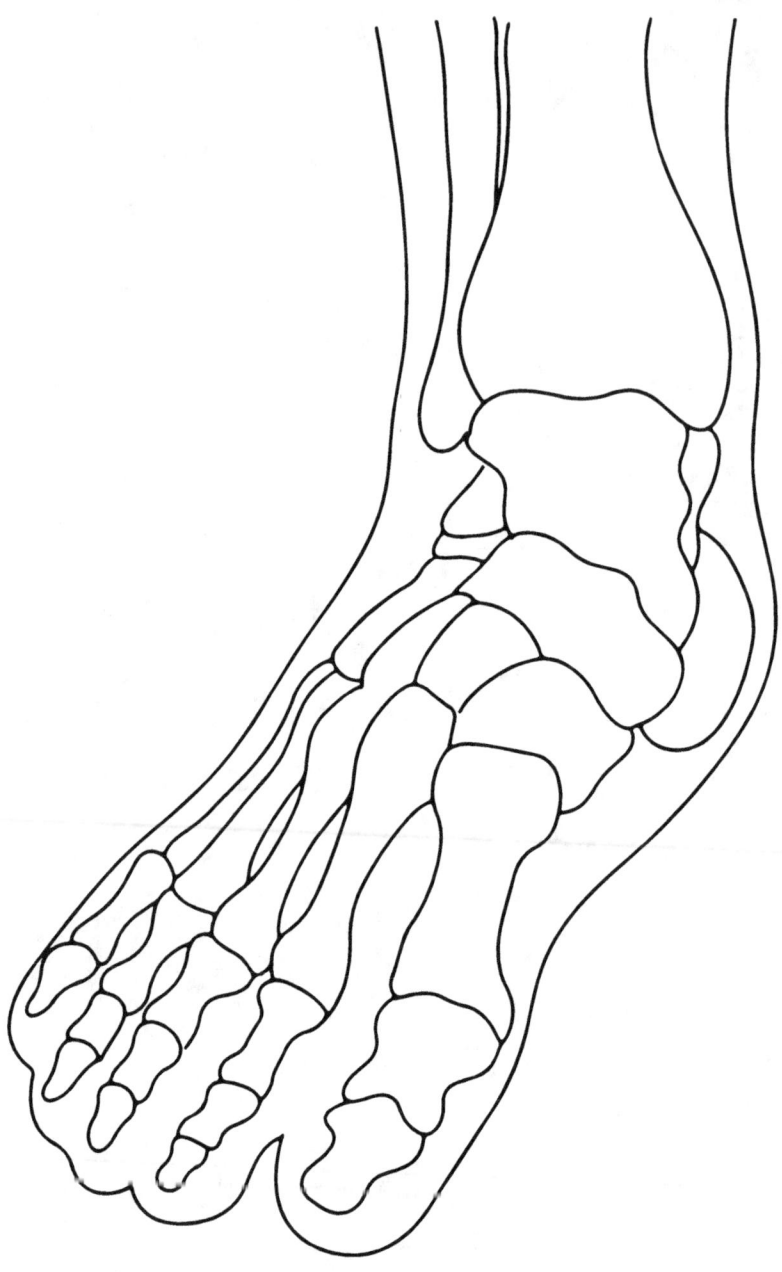

Leg

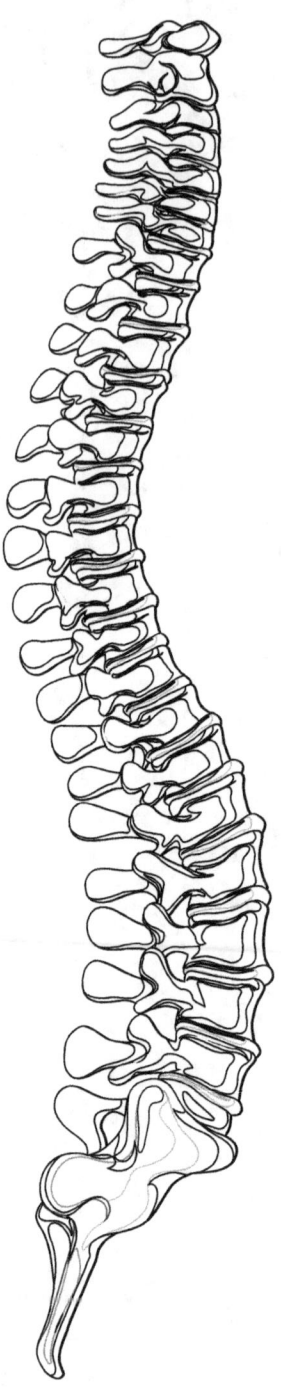

Spine

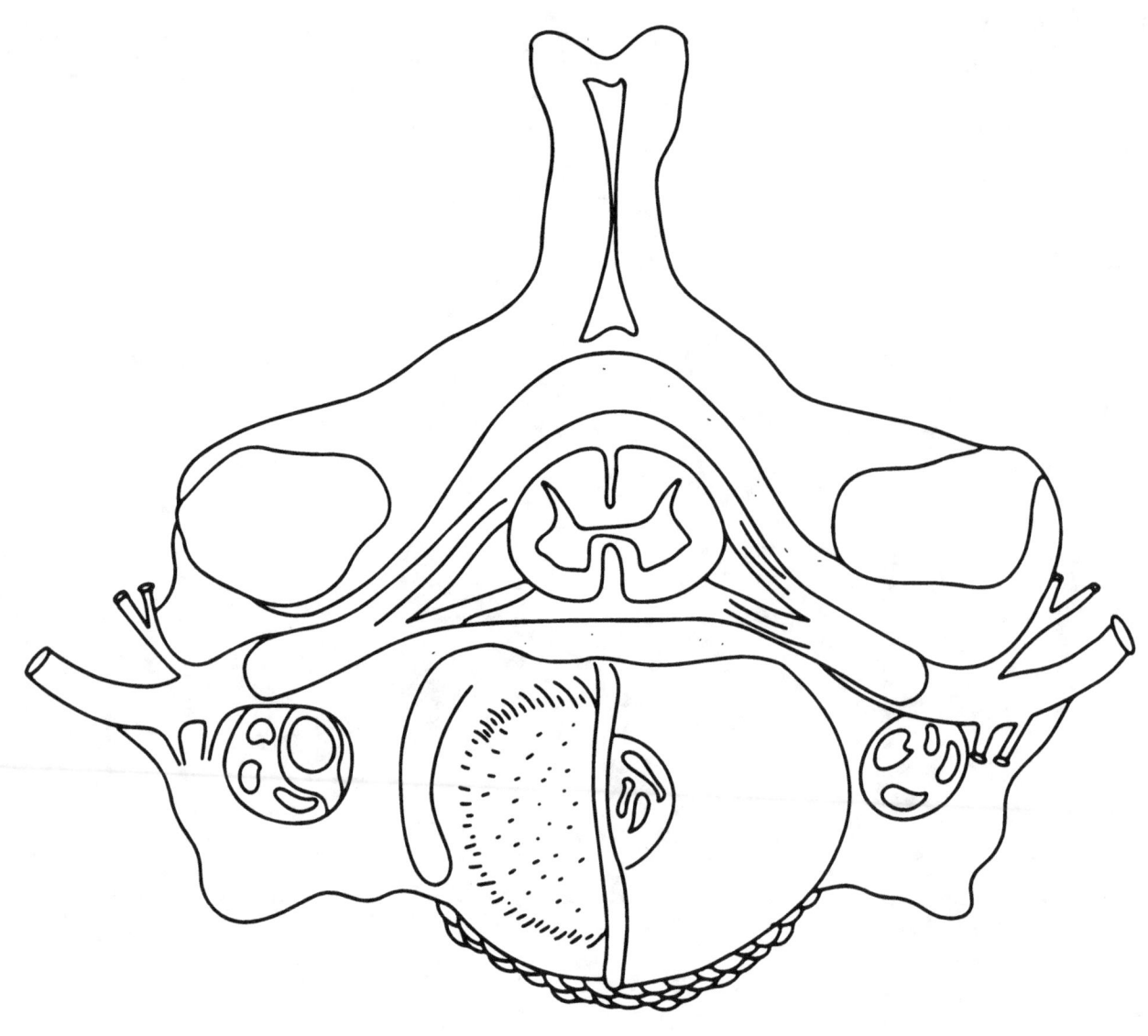

Spinal Cord

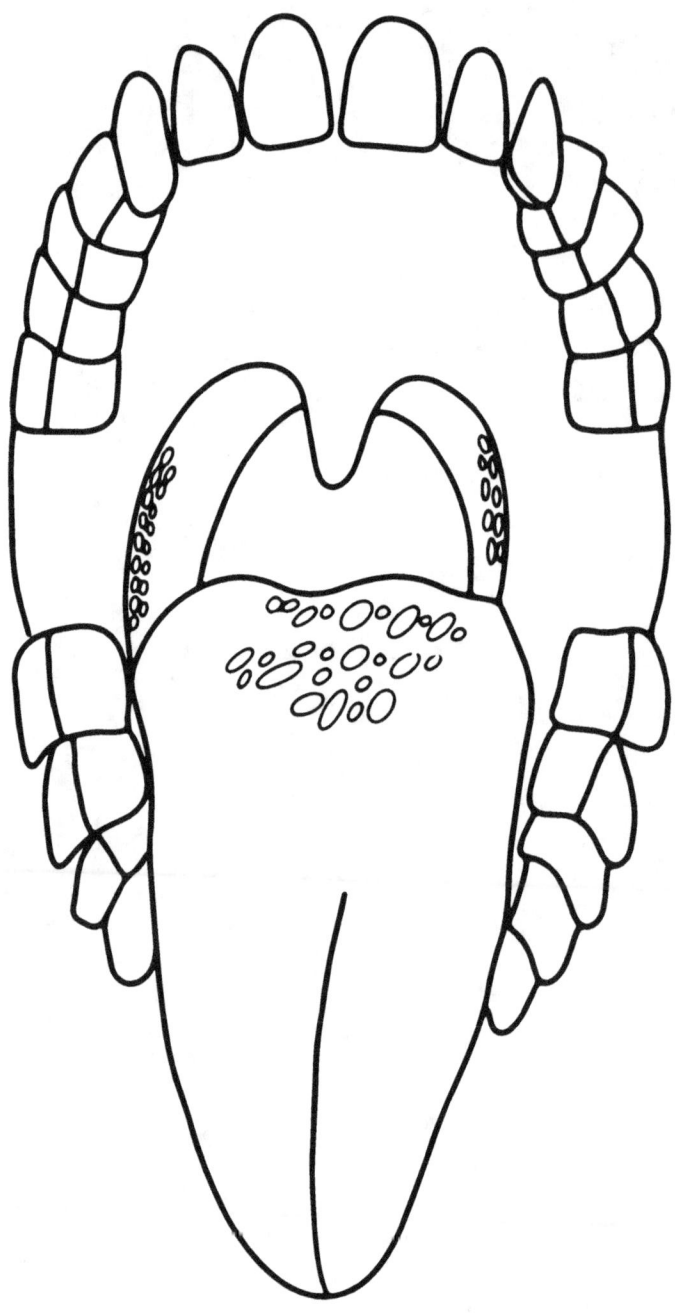

Tongue

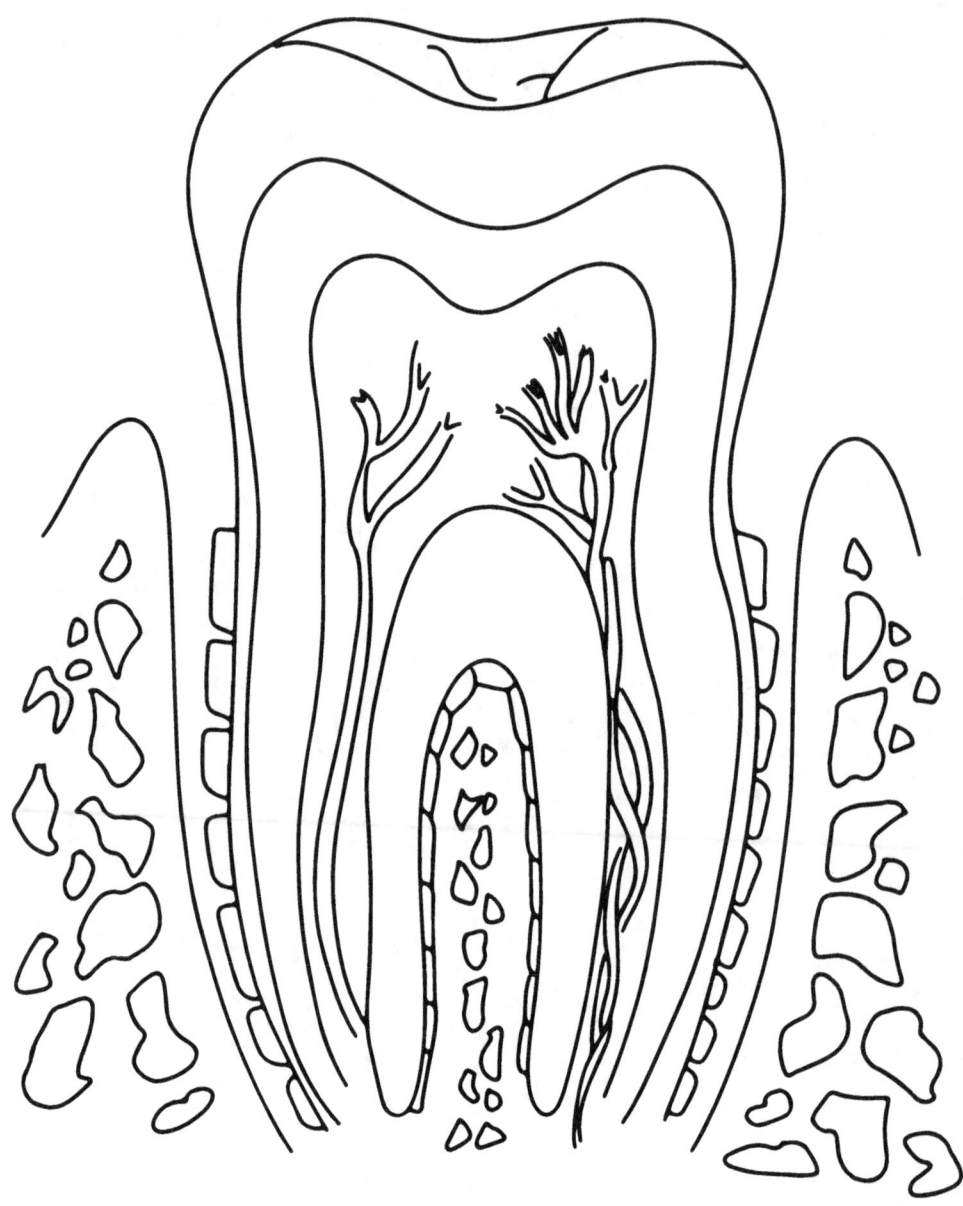

Teeth Structure

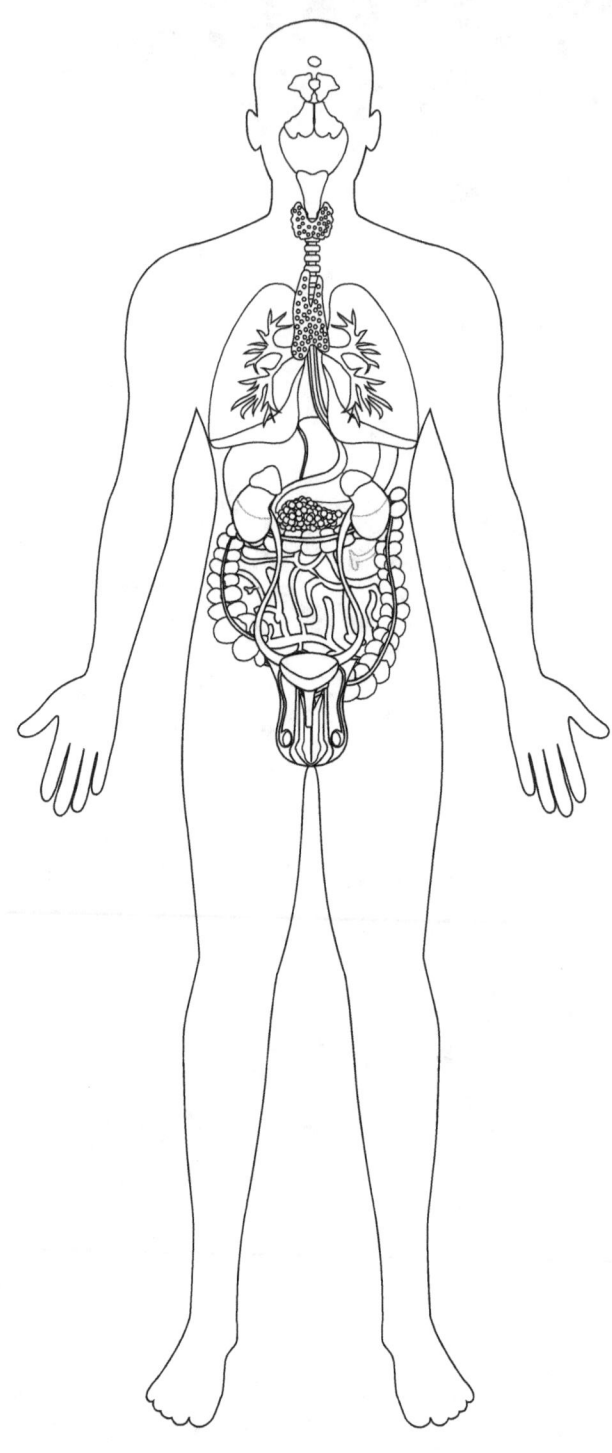

Internal Oragns

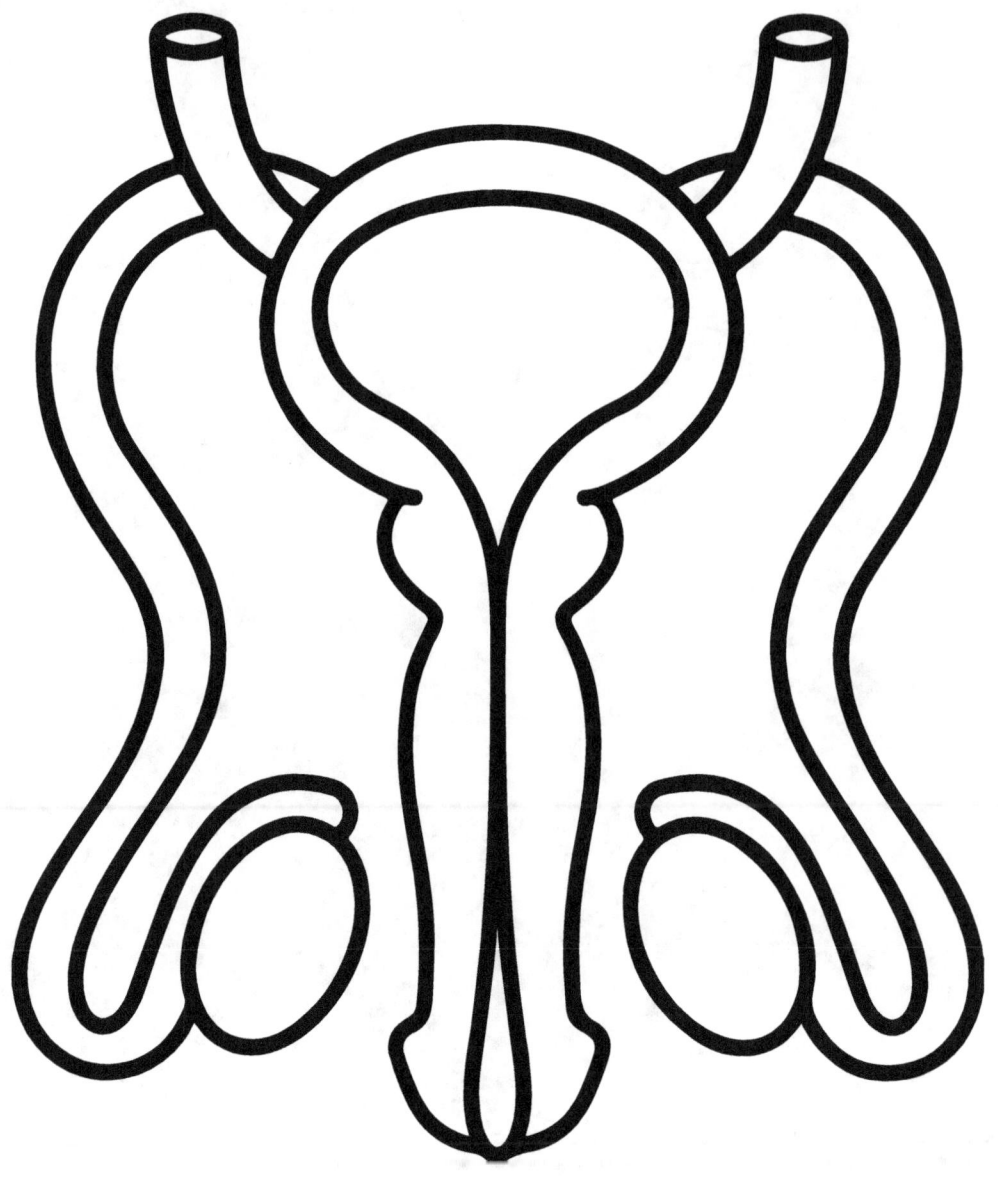

Male Reproductive System

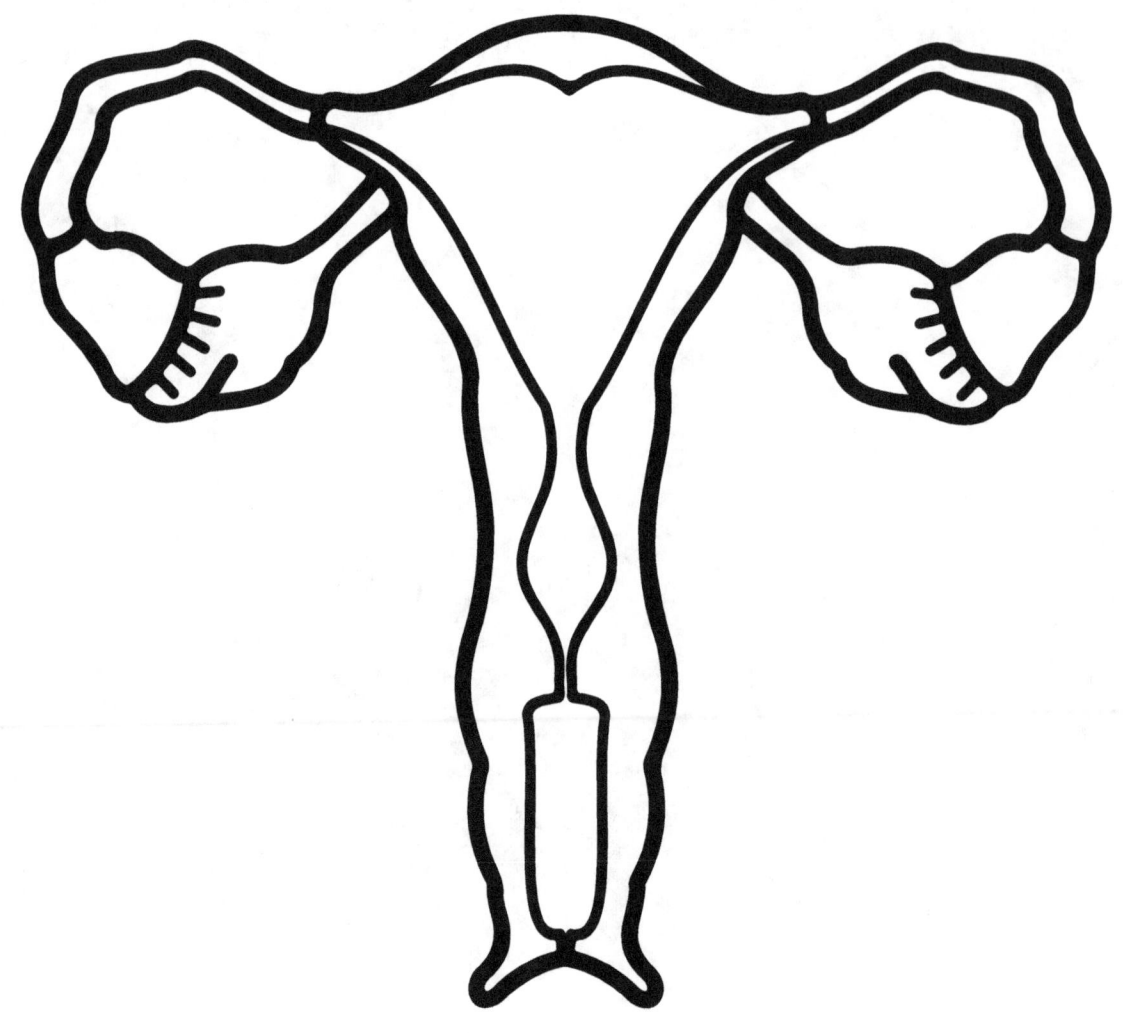

Female Reproductive System

Bladder

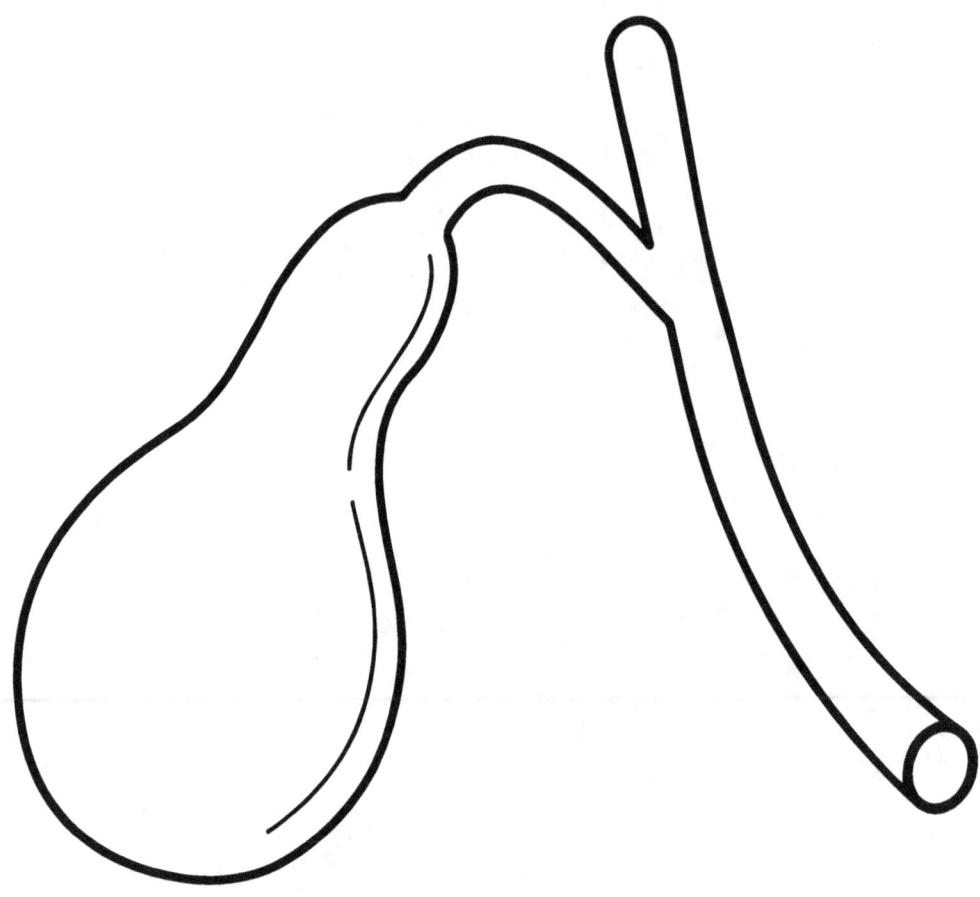

Gallbladder

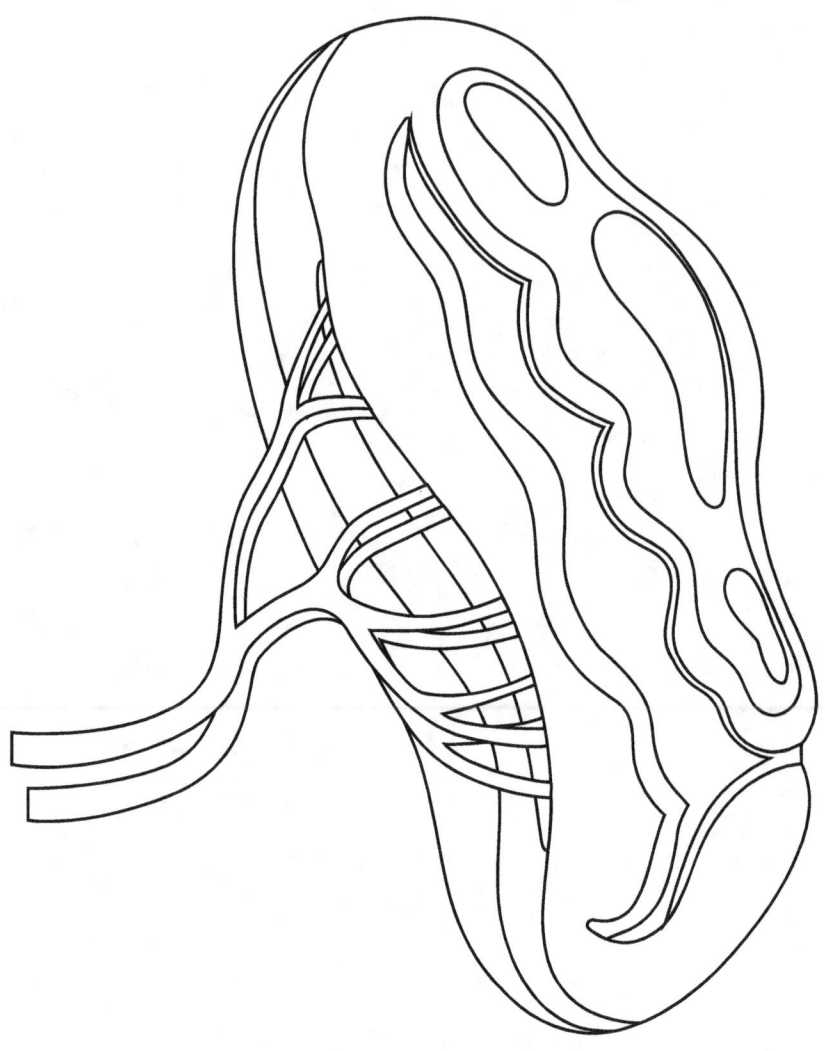

Spleen

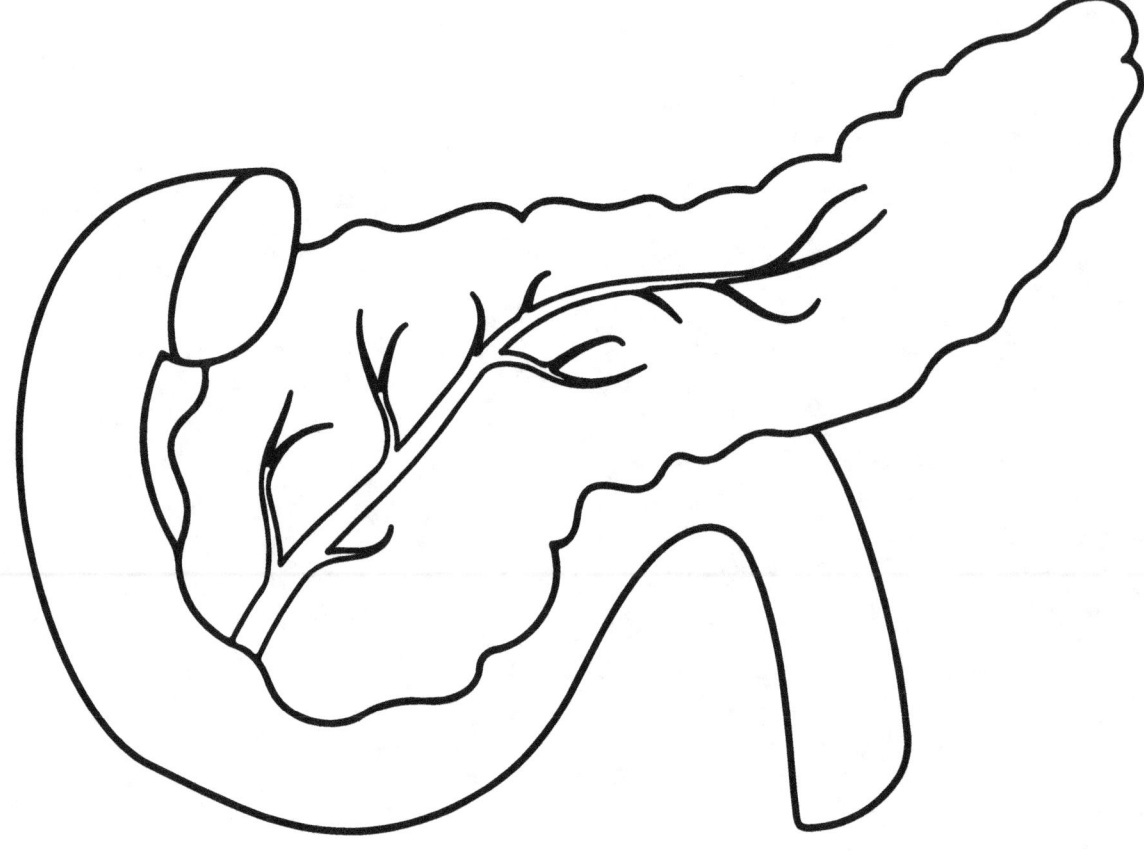

Pancreas

Skin

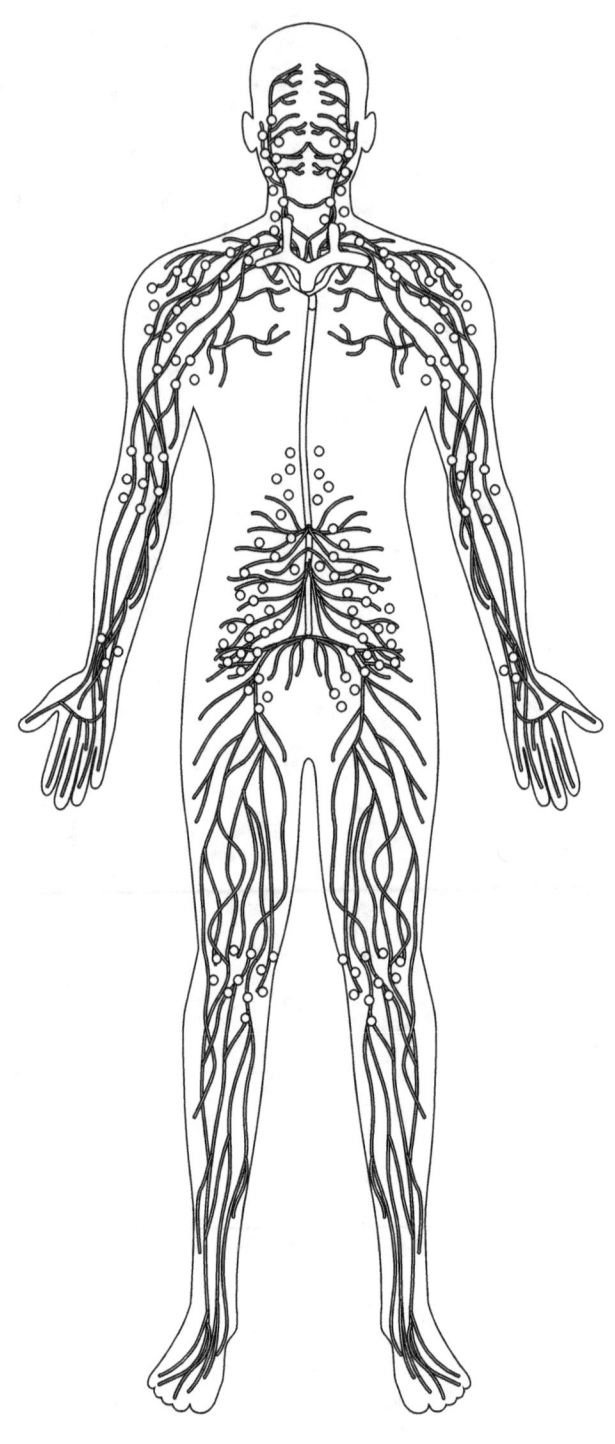

Lymphatic System

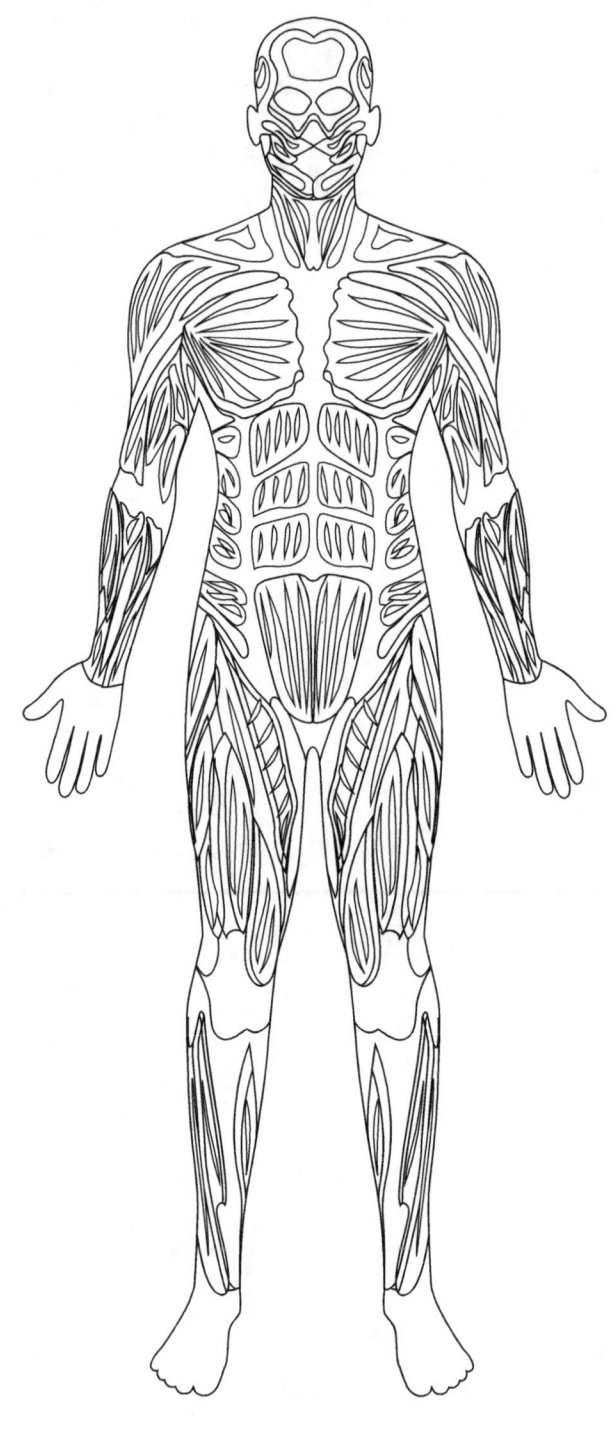

Muscle System

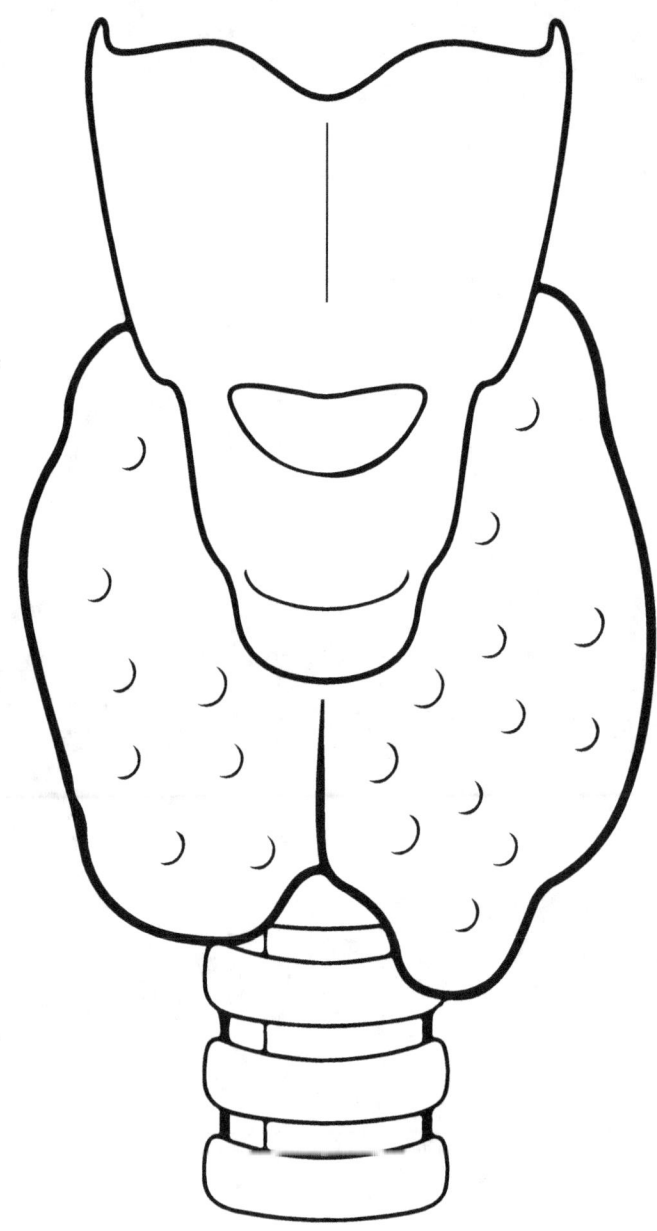

Thyroid

I Can See

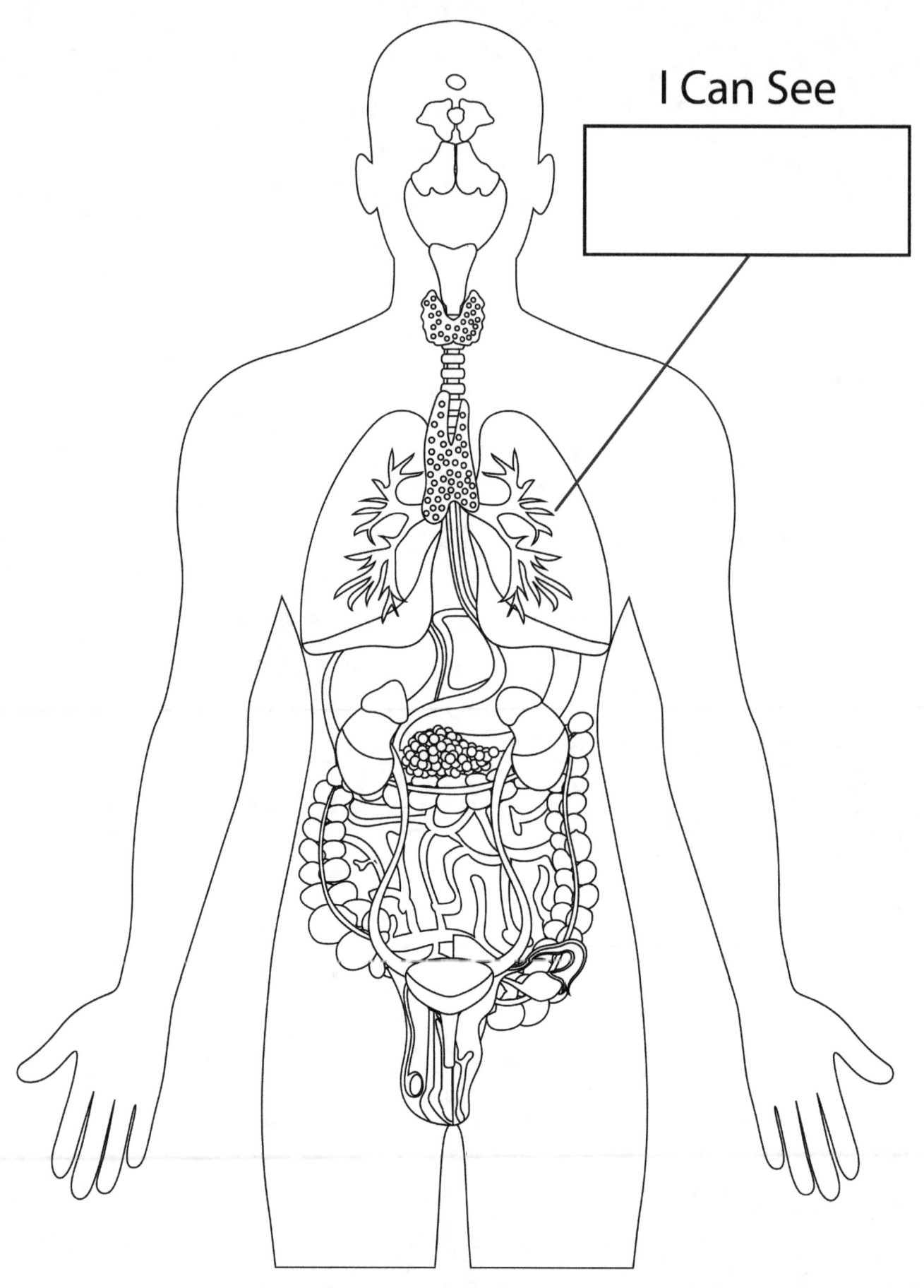

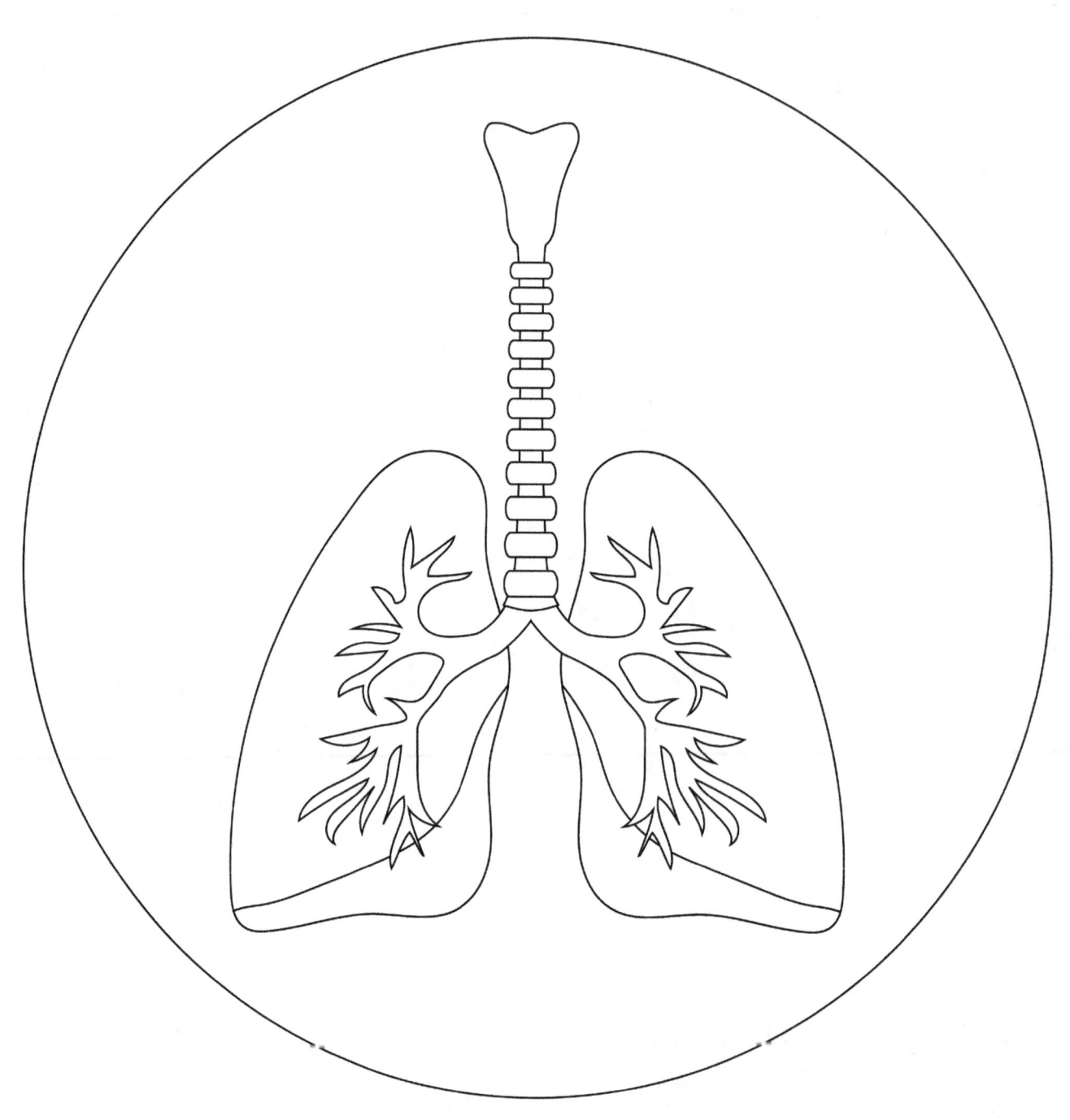

Lungs

I Can See

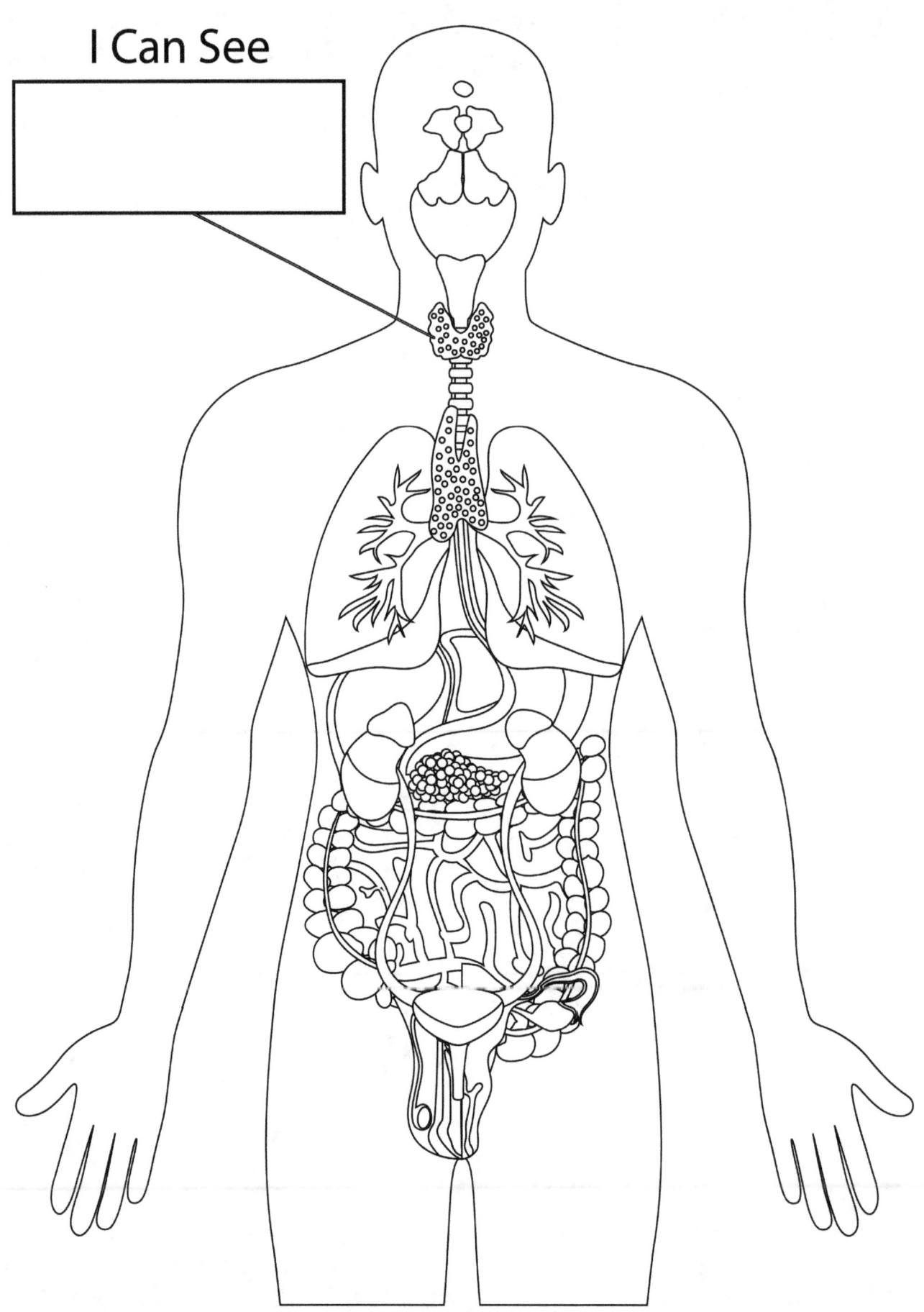

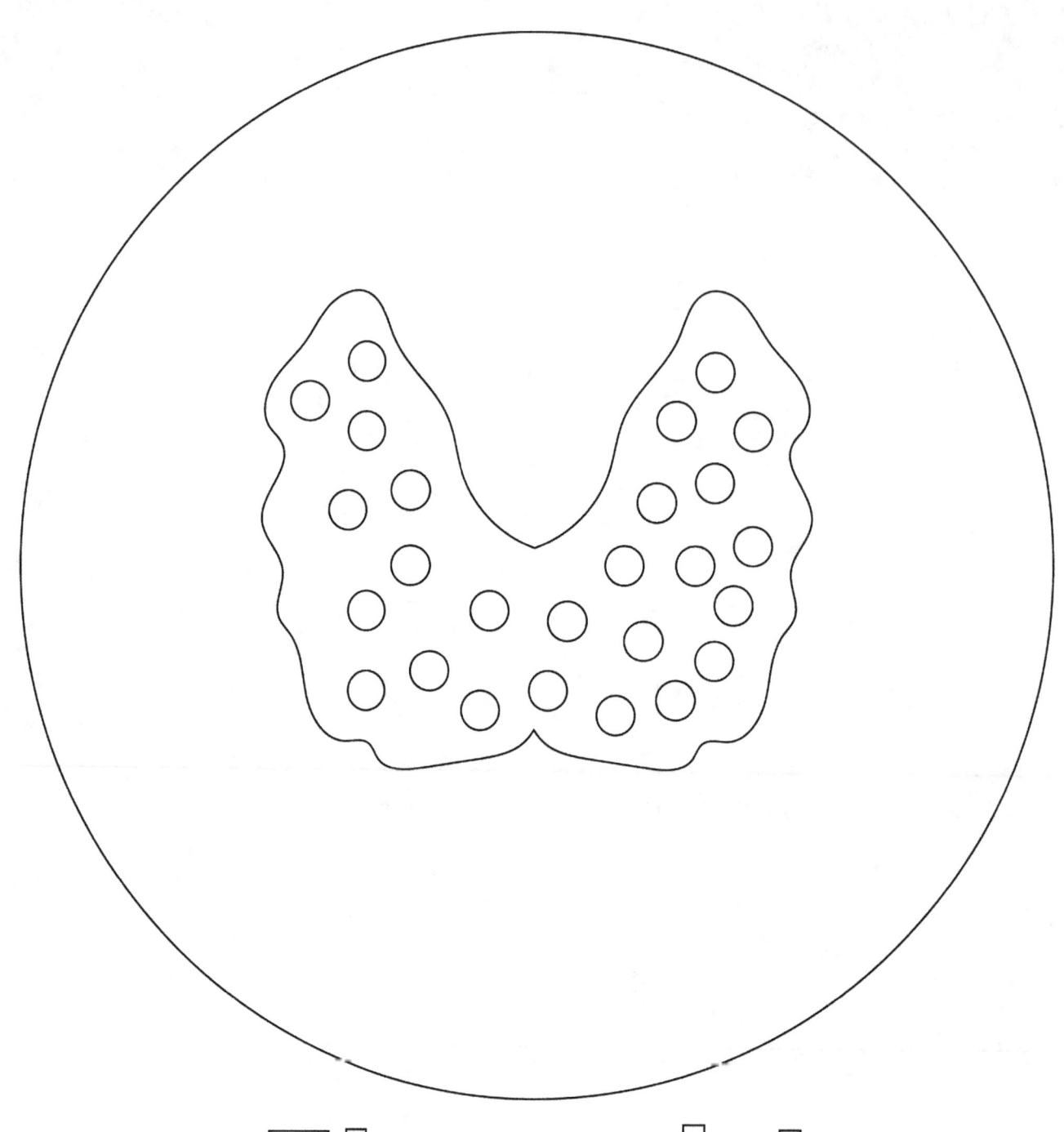

Thyroid

I Can See

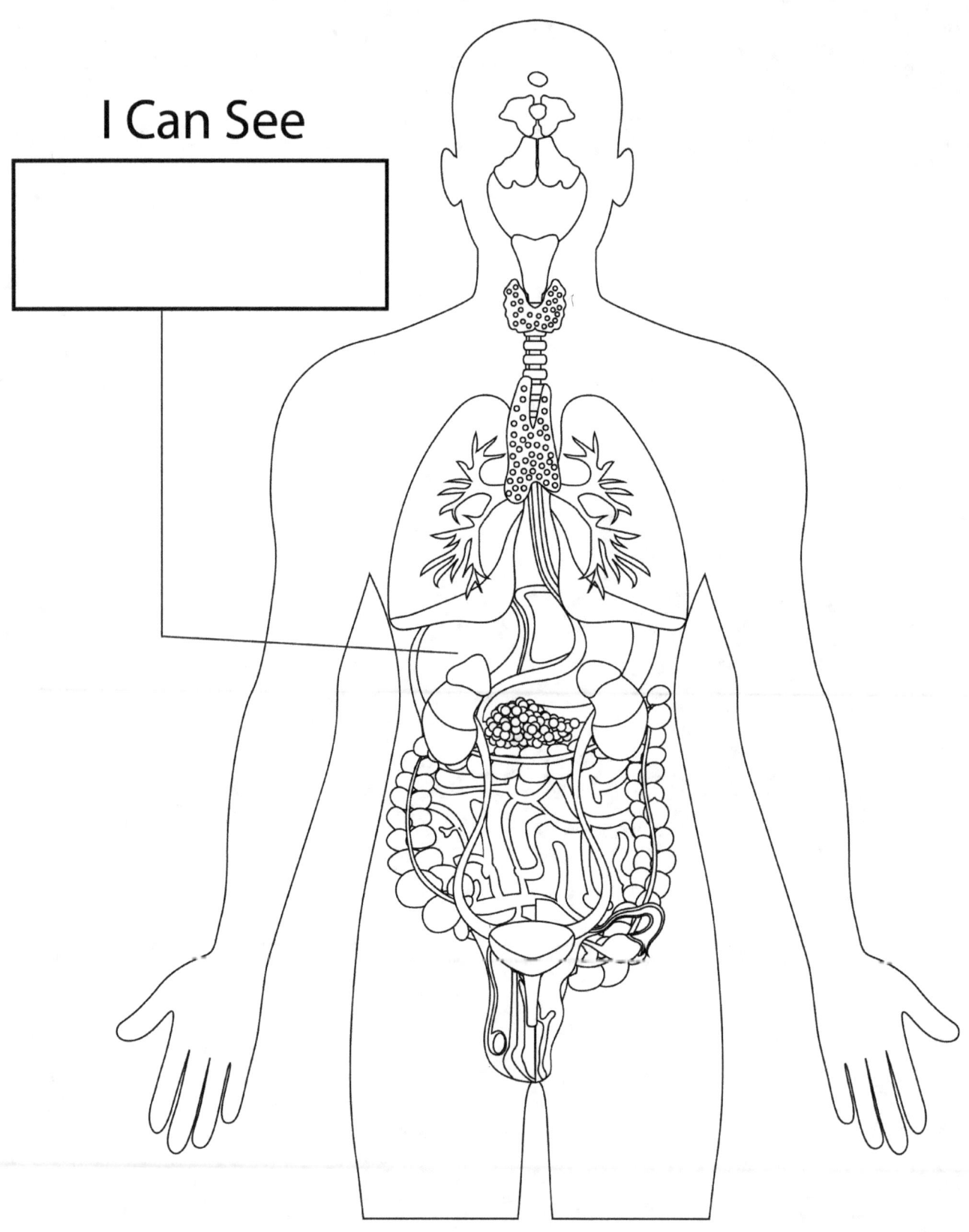

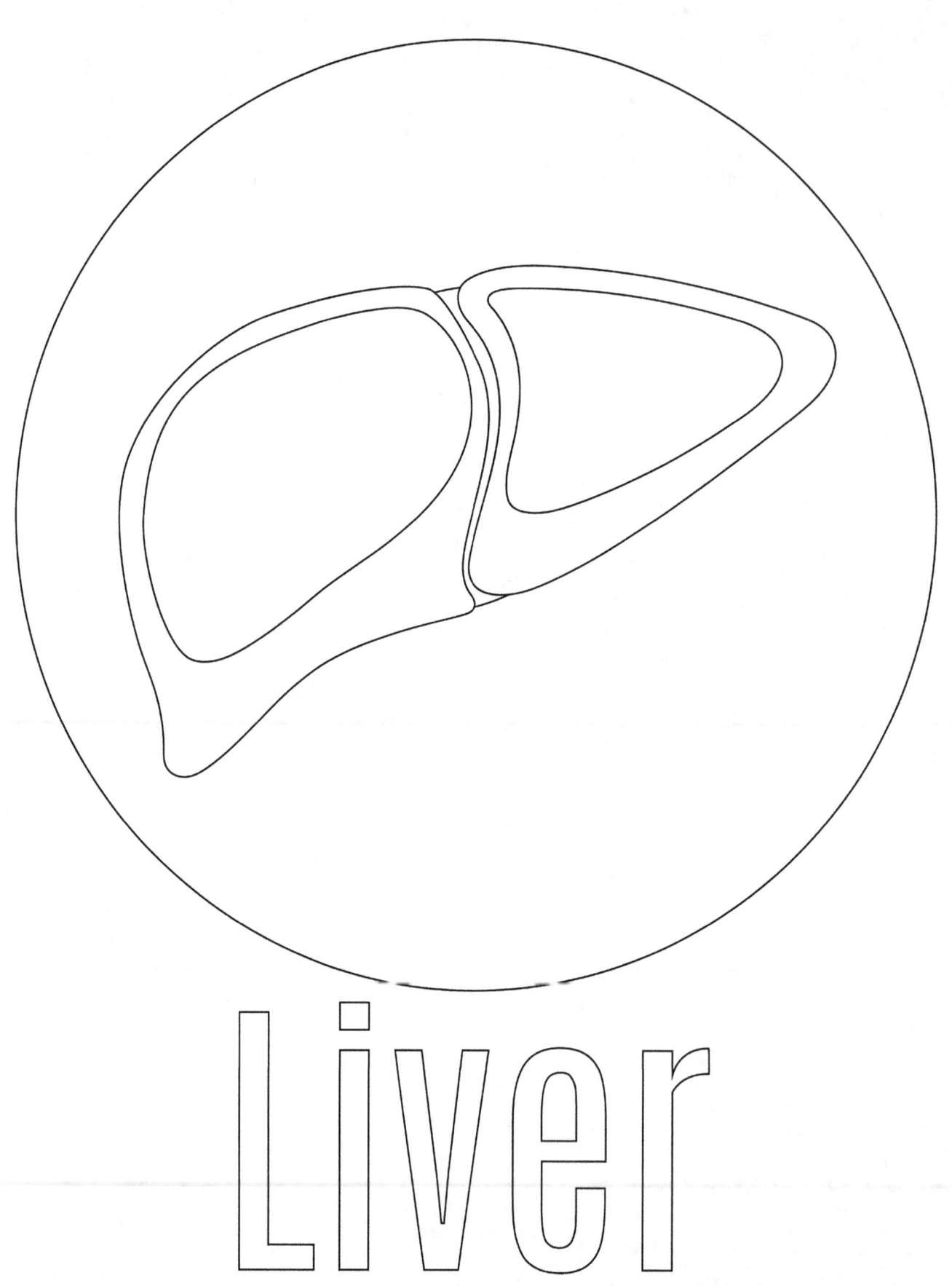

Liver

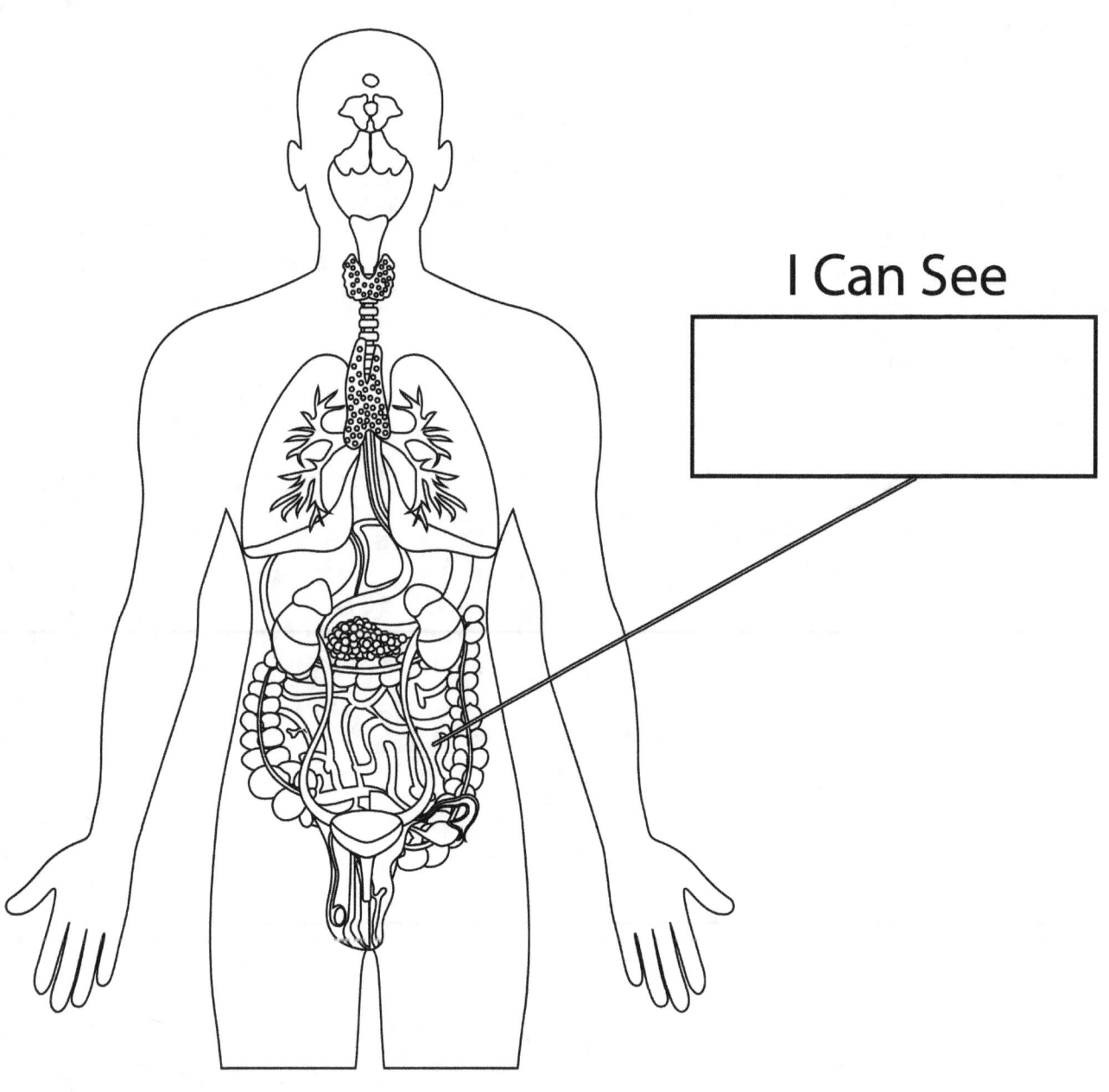

I Can See

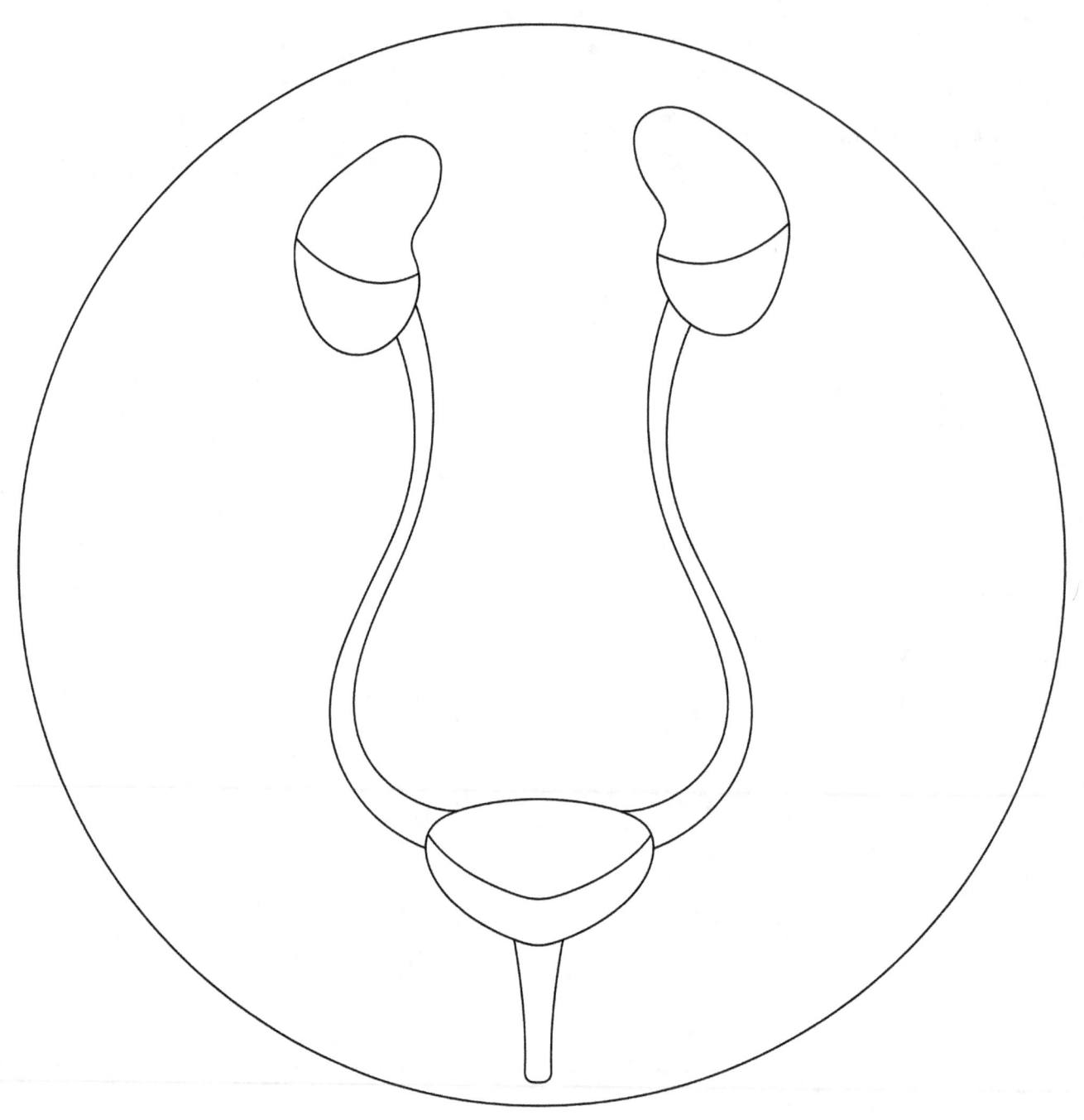

Urinary system

I Can See

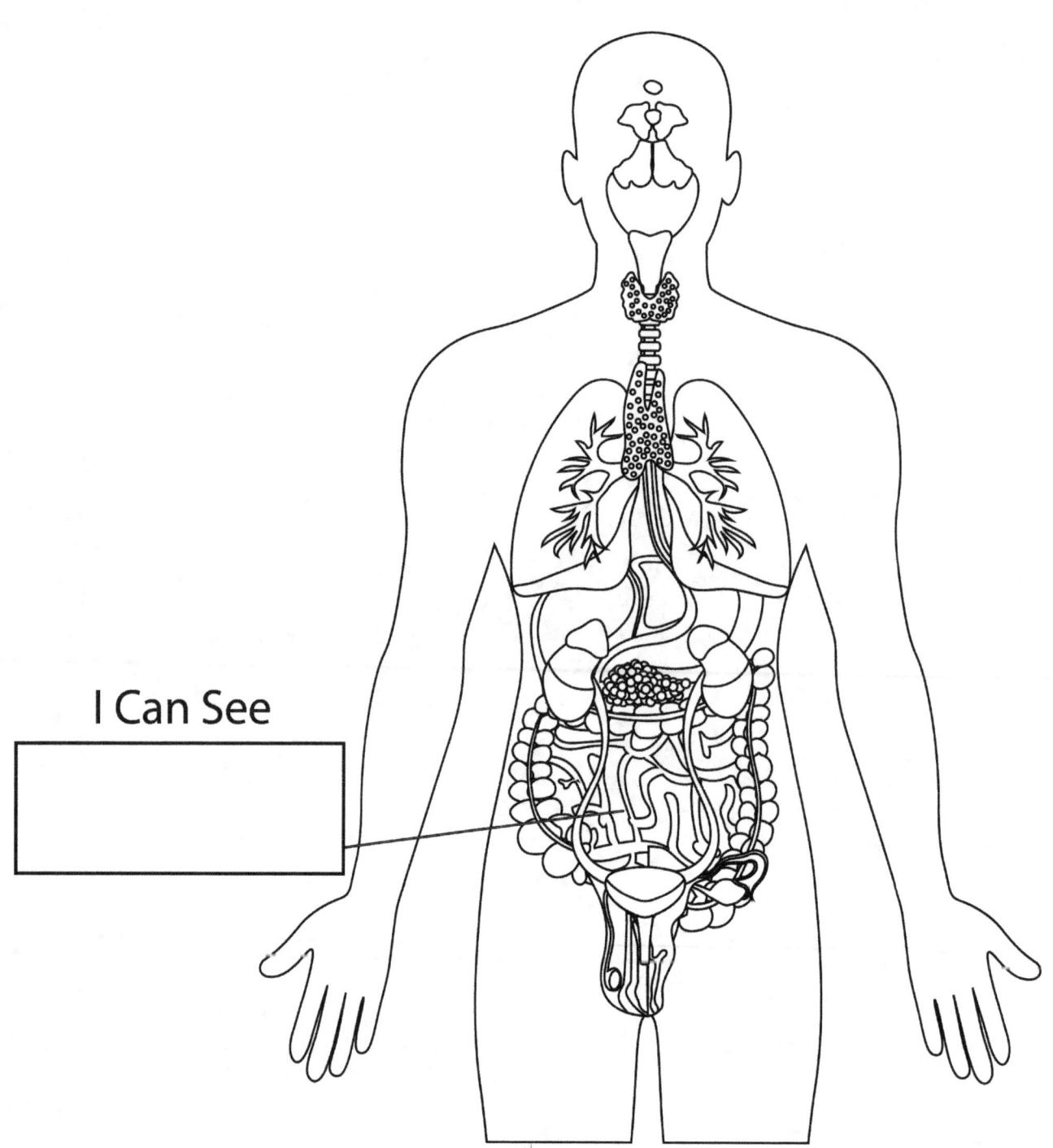

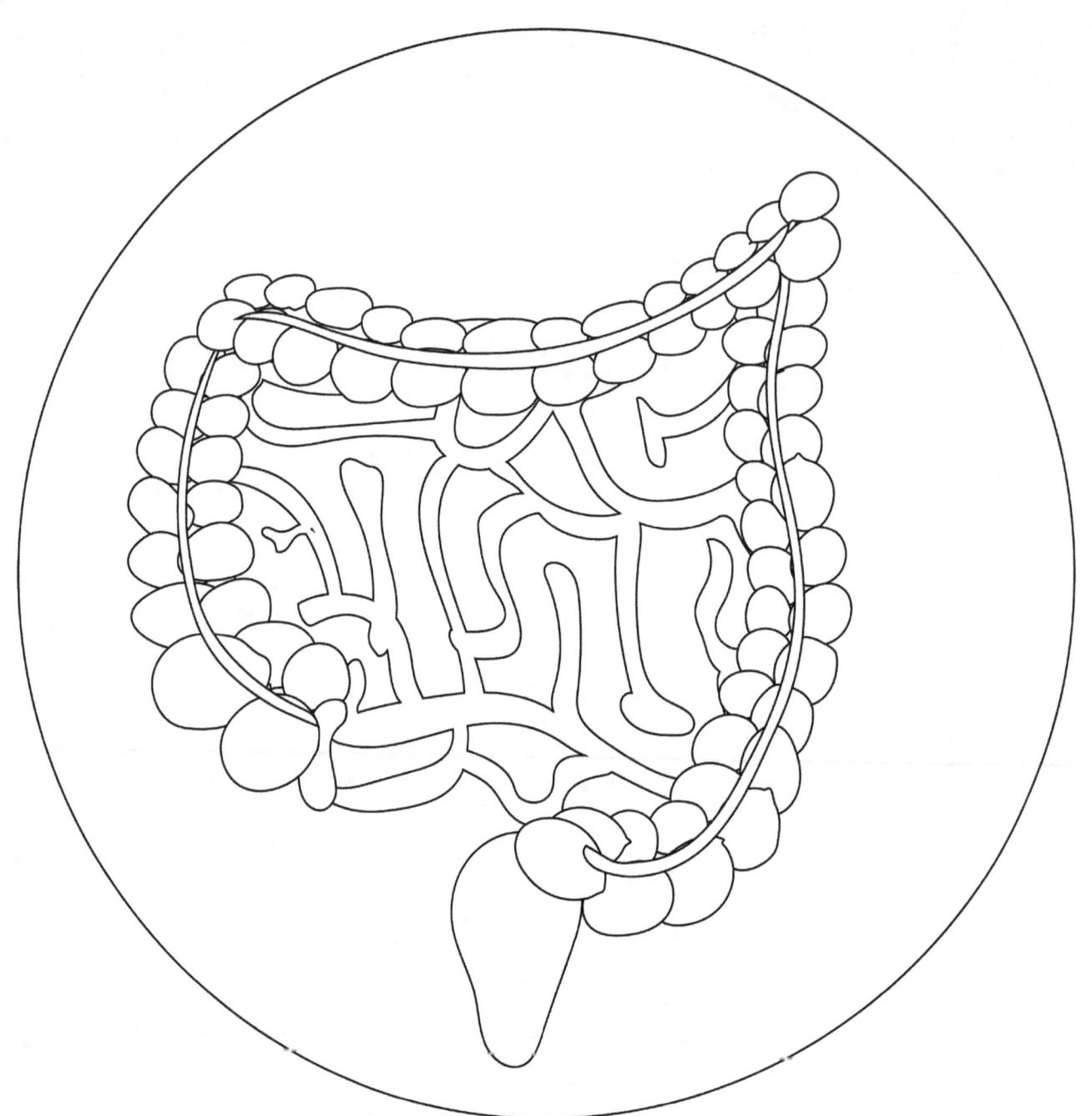

Intestines

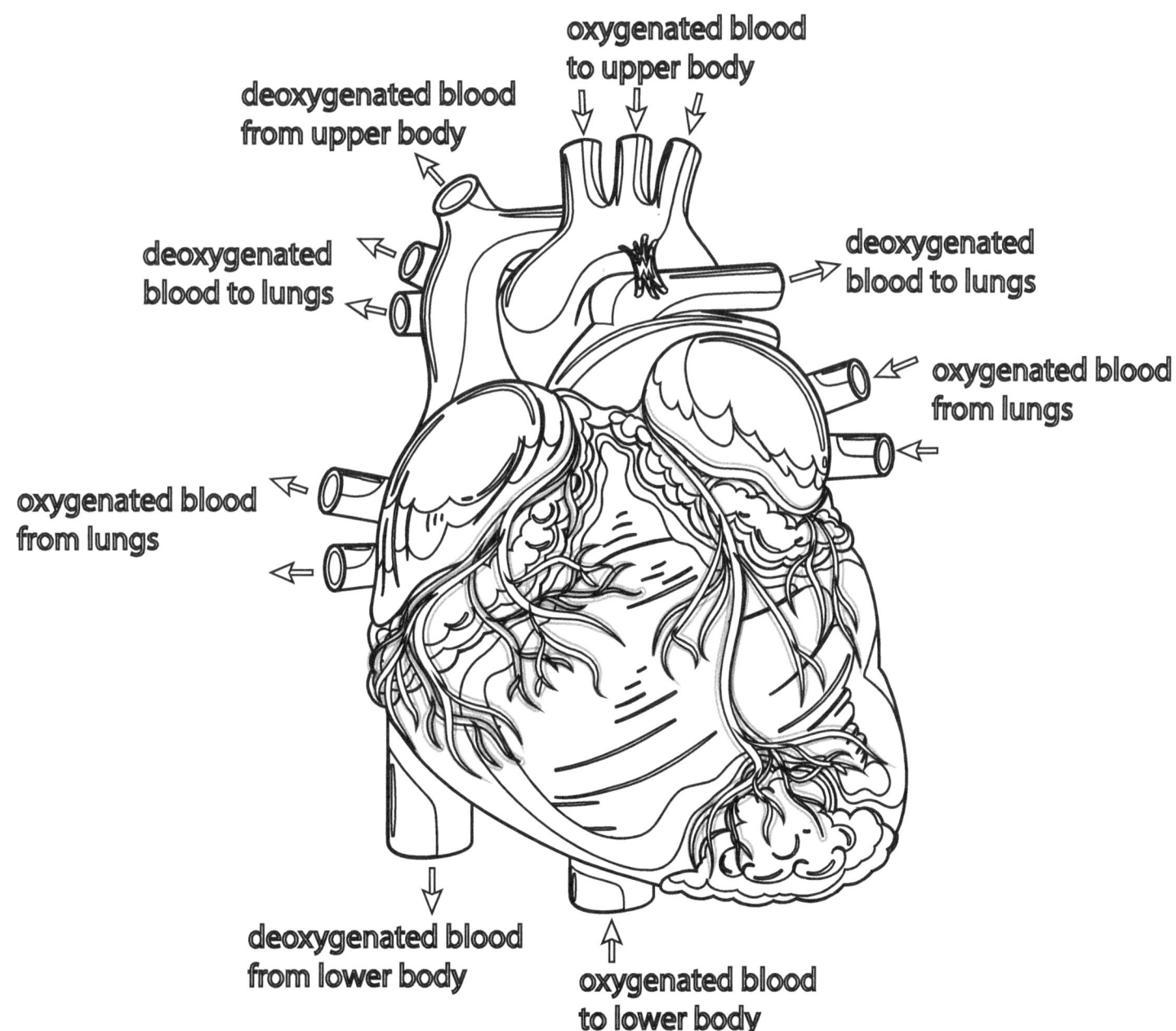

deoxygenated blood
from upper body

oxygenated blood
to upper body

deoxygenated
blood to lungs

deoxygenated
blood to lungs

oxygenated blood
from lungs

oxygenated blood
from lungs

deoxygenated blood
from lower body

oxygenated blood
to lower body